Security for Mobile Networks and Platforms

For a complete listing of the *Artech House Universal Personal Communications Series*, turn to the back of this book.

Security for Mobile Networks and Platforms

Selim Aissi
Nora Dabbous
Anand R. Prasad

**ARTECH
HOUSE**

BOSTON | LONDON
artechhouse.com

Library of Congress Cataloging-in-Publication Data
A catalog record for this book is available from the U.S. Library of Congress.

British Library Cataloguing in Publication Data
A catalogue record for this book is available from the British Library.

Cover design by Igor Valdman

Artech House
685 Canton Street
Norwood MA 02062

ISBN 10: 1-59693-008-x
ISBN 13: 978-1-59693-008-7

10 9 8 7 6 5 4 3 2 1

Contents

Preface

The idea of this book resulted from several discussions between the authors over the span of more than three years while representing their respective companies in the Third Generation Partnership Project (3GPP) and organizing conferences. They realized the lack of a single book that provided an end-to-end coverage of mobile and wireless security. While there are several excellent books that describe wireless network security, there is a clear need for a book that covers the three main components of mobile and wireless security: the network, the protocols, and the platforms. Since the authors' collective backgrounds cover all those aspects, the idea of the book turned into a plan that was quickly adopted by Artech House.

Selim Aissi is leading the manageability and security architecture of emerging mobile and wireless platforms at Intel, working on hardware, software, and protocol aspects of next-generation mobile platforms. Nora Dabbous is leading the architecture of several advanced cryptographic systems during her long tenure at Gemalto. Anand Prasad manages several next-generation wireless network security and architecture projects at DoCoMo. With this diverse wireless background, the three authors worked diligently for over two years to complete this book.

This book is intended for mobile and wireless professionals, managers, and students. It describes security principles and methods in a real-world context, discussing existing and emerging standards and technology trends. Because of the end-to-end security coverage, the book covers topics in moderate depth. For the reader who wants to delve deeper, each chapter includes extensive bibliographic notes. The reader needs no specific expertise, but familiarity with wireless and mobile concepts is helpful.

Wireless and mobile system architects and developers will learn the basic security principles and technologies that are relevant to their projects. Seniors and graduate students in computer science or information systems will find thorough coverage in a well-organized, readable form. We believe that all mobile and wireless professionals will find this book interesting and useful. For all other readers, this book can serve as a comprehensive reference and a guide to other literature.

The book has 16 chapters. Chapters 1 through 3 provide an introduction to mobile and wireless security concepts, including authentication, authorization, and cryptographic techniques.

Chapters 4 through 6 describe the security aspects of mobile platforms, including hardware and software security, and the important topic of mobile security certification. Chapters 7 through and 10 are focused on wireless protocols and describe higher layer and IP layer security, including network authentication and authorization from networking and protocol perspectives.

Chapters 11 through 14 describe network security aspects of wireless area networks: wireless personal area networks (WPANs), wireless local area networks (WLANs), wireless metropolitan area networks (WMANs), and wireless wide area networks (WWANs), respectively.

Chapter 15 provides some insights into emerging wireless and mobile technologies. Chapter 16 provides a catalog of mobile security threats.

Figures 12.3–12.7 are reprinted with permission from IEEE std. 802.11i—2004 Amendment to IEEE std. 802.11, 1999 Edition (Reaff, 2003). IEEE Standard for Information Technology—Telecommunications and information exchange between systems—Local and metropolitan area network-specific requirements—Part II: Wireless LAN Medium Access Control (MAC) and Physical Layer (PHY) specifications—Amendment 6: Medium Access Control (MAC) Security Enhancements.

1

Introduction

1.1 Overview

This section of the first chapter will provide an introduction to security to the reader. The chapter will start with basic information on what is meant by security in general for telecommunication systems. The section will also provide an introduction to security threats and the five mantras of security services. Common solutions for security issues that make use of various security services will discussed in this section. After the basics, security in wireless and Internet protocol (IP) will also be introduced.

1.1.1 Security Basics

Security always takes backstage when it comes to the functionality of any product, although the success of any product is dependent on it. Any attack on a system leads to mistrust by customers and thus, a decrease in business. Some of the issues are well known. Some say that security is the holy grail of telecommunication systems; G. Spafford says [1]: The only truly secure system is one that is powered off, cast in a block of concrete, and sealed in a lead-lined room with armed guards—and even then I have my doubts.

It is important to understand the definition of security first. The definition given by ISO/IEC 2382-8 [2] is *the protection of data and resources from accidental or malicious acts, usually by taking appropriate actions.*

This book discusses security in various telecommunications systems, covering solutions and issues. Telecommunications for us mainly refers to wireless systems that have the inherent security issue of being open to all.

1

One of the most prominent mistakes in the design of any system is the consideration that security is a feature that can be added. This is a flawed way of thinking that often only leads to several subsequent issues in a later phase. The issues can lead to a loss of billions of dollars or simply a loss of trust in a company or product. A simple example is 802.11-based WLANs, where the design of medium access control (MAC) can lead to denial of service (DoS) attacks. Though new security solutions (e.g., 802.11i) may counteract some existing attacks, the issue remains in terms of insecure management frames that can lead to man-in-the-middle attacks. It is extremely important to consider the complete system and look at the business and usage scenarios to determine the security vulnerabilities.

To implement security, a few steps are generally taken:

1. Determine the assets.
2. Determine the threats and risks to each asset and thereby set security requirements.
3. Design and implement countermeasures for the threats and residual risks to an economic level.
4. Monitor, manage, and update the implementation.
5. Finally, deter, detect, and react against any attacks.

These steps lead to security solutions during the complete product lifetime.

Before proceeding further, let us look at some definitions [3]:

- *Asset:* Anything that is of value.
- *Vulnerability:* Any weakness that could be exploited to violate the security of a system or the information that it contains.
- *Threat:* A potential violation of security; it is either accidental (e.g., program bug) or intentional (e.g., hacking), or it is active (e.g., unauthorized data modification) or passive (e.g., wiretapping).
- *Attack:* Realization of threat, successful or not; it is either active or passive;
- *Intrusion:* Successful attack.
- *Risk:* Potential that a given threat will exploit vulnerabilities to cause damage to assets.
- *Risk management* or sometimes *security management:* the balancing of the appropriate actions to be taken in order to protect the organization to an appropriate level.

- *Countermeasures* or *safeguards:* Mechanisms used to protect assets from harm or decrease the effects of intrusion.

- *Residual risks:* Risks that are accepted and are not planned to be mitigated due economic or other considerations.

1.1.2 Threats and Attacks

The introduction of distributed systems and the use of networks and communications facilities—wireline and now increasingly wireless—have increased the need for network security measures to protect data—both in real time and non-real time—during transmission. To effectively assess the security needs, and evaluate/choose the most effective solution, a systematic definition of the security goals or requirements and an understanding of the threats is a necessity. In this section, first the security threats and then the security goals are discussed. This section also discusses which security goals will counter a given security threat.

1.1.2.1 Threats

Security threats or security issues can be divided into two types: passive and active threats. Passive threats stem from individuals attempting to gain information that can be used for their benefit or to perform active attacks at a later time. Active threats are those where the intruder does some modification to the data, network, or traffic in the network. In the following section, the most common passive and active threats are discussed.

1.1.2.2 Passive Threats

A passive threat is a situation in which an intruder does not do anything to the network or traffic under attack but collects information for personal benefit or for future attack purposes. Two basic passive threats are described as follows:

- *Eavesdropping:* This has been a common security threat to human beings for ages. In this attack, the intruder listens to things he or she is not supposed to hear. This information could contain, for example, the session key used for encrypting data during the session. This kind of attack means that the intruder can get information that is meant to be strictly confidential.

- *Traffic analysis:* This is a subtle form of passive attack. At times, it is enough for the intruder to know simply the location and identity of the communicating device or user. An intruder might require only information such as a message has been sent, or who is sending the message to whom, or the frequency or size of the message. Such a threat is known as traffic analysis.

1.1.2.3   Active Threats

An active threat arises when an intruder directly attacks the traffic and the network and causes a modification of the network, data, and so forth. The following list details common active attacks:

- *Masquerade:* This is an attack in which an intruder pretends to be a trusted user. Such an attack is possible if the intruder captures information about the user, such as the authentication data simply the username and the password. Sometimes the term *spoofing* is used for masquerade.

- *Authorization violation:* This occurs when an intruder or even a trusted user uses a service or resources that is not intended for that user. In the case of an intruder, this threat is similar to masquerading; having entered the network, the intruder can access services he or she is not authorized to access. On the other hand, a trusted user can also try to access unauthorized services or resources; this could be done by the user performing active attacks on the network or simply by lack of security in the network/system.

- *DoS:* DoS attacks are performed to prevent or inhibit normal use of communications facilities. In the case of wireless communications, it could be as simple as causing interference, sending data to a device and overloading the central processing unit (CPU), or draining the battery. Such attacks could also be performed on a network by, for example, flooding the network with unwanted traffic.

- *Sabotage:* A form of DoS attack, that could also mean the destruction of the system itself.

- *Modification or forgery of information:* This occurs when an intruder creates new information in the name of a legitimate user or modifies or destroys the information being sent. It could also be that the intruder simply delays the information being sent. An example is an original message Allow Alice to read confidential Source Codes modified to Allow Bob to read confidential Source Codes.

1.1.3   Security Services

Five major security goals, also known as security services, can be used as *security requirements.* These goals are discussed as follows:

- *Confidentiality:* This is for the protection of data from disclosure to an unauthorized person. Encryption is used to fulfill this goal. With an active attack, it is possible to decrypt any form of encrypted data (given there is a good mathematician/cryptographer or a person with a powerful computer and no time limit); thus, confidentiality is primarily considered a protection against passive attacks.

- *Authentication:* The authentication service is concerned with assuring that a communication is authentic. In the case of a single message, such as a warning or alarm signal, the function of the authentication service is to assure the recipient that the message is from the source that it claims to be from. In the case of an ongoing interaction, such as the connection of a terminal to a host, two aspects are involved. First, at the time of connection initiation, the service assures that the two entities are authentic (i.e., each is the entity that it claims to be). Second, the service must assure that the connection is not interfered with in such a way that a third party can masquerade as one of the two legitimate parties for the purposes of unauthorized transmission or reception.

- *Access control:* In the context of network security, access control is the ability to limit and control access to the systems, the networks, and the applications. Thus, unauthorized users are kept out. Although given separately, user authentication is often combined with access control purposes; this is done because a user must be first authenticated (e.g., by the given server and the network) so as to determine the user access rights. Implicitly, access control also means authorization.

- *Integrity:* This prevents unauthorized changes to the data. Only authorized parties are able to modify the data. Modification includes changing status, deleting, creating, and delaying or replaying of the transmitted messages.

- *Non-repudiation:* Neither the originator nor the receiver of the communication should be able to deny the communication and content of the message later. Thus, when a message is sent, the receiver can prove that the message was in fact sent by the alleged sender. Similarly, when a message is received, the sender can prove that the message was in fact received by the alleged receiver.

Besides these security requirements, some general requirements play an important role in developing the security solutions; these are the following:

- *Manageability:* The load of the network administrator must not be unnecessarily increased by adding security, while the deployed solutions should be easy to manage and operate in the long term.

- *Scalability:* A network must be scalable, which requires the security scheme deployed in the network to be equally scalable while maintaining the level of security. Here the term scalability is being used in its broadest sensescalable in terms of the number of users and in terms of an increase in network size (i.e., addition of new network elements or an extension to a new building).

- *Implementability:* A simple and easy way to implement a scheme is extremely important. Thus, a security scheme must be devised so that it is easy to implement and still fulfills the security requirements.

- *Performance:* Security features must have minimum impact on the network performance. This is especially important for real-time communication, where the security requirements must be met while the required quality of service is met. Performance also goes hand in hand with the resource usage of the medium; the security solutions must not, for example, cause a decrease in the overall capacity of the network.

- *Availability:* This goal is closure to the five goals mentioned earlier in the section. Any service or network should be available to the user. Several attacks are possible to disrupt the availability, with DoS being the major one.

1.1.4   Common Security Solutions

In this section, the security threats are mapped to the security goals. Upon knowing which threat can be countered by which goal, the next step is to find the security solutions or mechanisms that can fulfill the security goals. A mapping of security threats and goals is given in Table 1.1 [1], where X in a cell indicates that the given security goal can counter the given threat. It should be noted that a security goal can sometimes fail to counter a threat.

In wireless systems, one of course has to observe that the communication takes place on a medium that is accessible to all. On the other hand, the basic issue of IP-based communications is that IP is well understood by many, and the Internet has provided access to tools that can be used even by a novice to successfully perform attacks.

1.2   Trusted Mobile Platforms

1.2.1   Introduction

The mobile industry depends on the deployment of new, attractive, and rich data services such as mobile commerce (mCommerce) and media services to increase its average revenue per unit (ARPU) and create new revenue sources. The successful deployment of these emerging services must be done without disruptions and outages of the wireless networks, or platform-level attacks that

Table 1.1
Mapping of the Security Threats to the Security Goals

Security Goals	Security Threats					
	Eaves-dropping	Traffic Analysis	Masquerade	Authorization Violation	DoS	Modification
Confidentiality	X	X	X	X		
Authentication			X	X		X
Access control			X	X		X
Integrity			X	X		X
Non-repudia-tion			X	X		X
Availability			X	X	X	X

could jeopardize the user's private information and data stored on the mobile platform. With these emerging mobile services and applications, users need to protect their data from street crime and cyber crime, content owners and providers need to be able to protect and charge for their content and services, and service providers need to be able to protect their networks against malicious use.

The security vulnerabilities of existing mobile and wireless platforms constitute a substantial risk to existing networks. In fact, over the past few years, the mobile and cellular industries have experienced an alarming increase in the number of handset thefts and associated fraud. There has also been an increase in the number of attacks by hackers against cellular devices. The U.S. Secret Service's Financial Crimes Division [4] has estimated the telecommunication fraud losses at more than a billion dollars yearly and that mobile phone cloning is one of the largest threats.

Also, some emerging technology trends are creating new threats to the security of mobile platforms and networks. Emerging Smartphones are optimized primarily for data services that primarily require that the device becomes an open platform for software applications. While this is essential to deliver a wide range of new applications and services, it also leads to mobile platforms that are more vulnerable to attack.

With the emergence of the Cabir virus and several of its derivatives, there's a great potential to see a rapid propagation of viruses over mobile networks as well as DoS attacks on the operator's services. With these attacks, which have been historically tied to the desktop personal computers (PCs) and fixed networks, lies a greater threat to user's private data stored on mobile platforms, including private keys used for financial transactions, email applications, and even remote access to corporate networks. There's even a threat to potentially disrupt the normal operation of the mobile platforms, such as blocking or redirecting cellular calls. There's also an opportunity for fraud to be committed on a large scale by breaking the international mobile equipment identity (IMEI) on GSM-enabled mobile platforms.

As for the proliferating enterprise usages of mobile devices, the adoption of such devices will be limited by the ability to demonstrate protection for company assets. The use of mobile devices in the corporate environment brings a new range of vulnerabilities.

The following section explains what it takes to develop a trusted mobile platform. In this book, the terms trusted, trustworthy, and secure are often used to indicate the same property: worthy of trust by the user and the service-providing party. There's no attempt in the book to distinguish between the historical meanings of such properties, and the reader is referred to a wealth of academic and industry publications and debates about the topic of trust. The authors believe that, although the importance of trust has long been recognized as paramount for the development of secure and dependable platforms, the meaning associated with trust or a trusted principal is seldom clearly defined. The fragmentation in trust-related naming conventions is partially caused by the lack of security standards for mobile platforms. The main goals of this book are to explain the security aspects of mobile platforms, networks, and protocols that drive the use of mobile platforms as well as to explain existing and emerging physical protections that protect mobile platforms from malicious actions.

1.2.2 Trusted Mobile Platforms

A trusted mobile platform is designed to establish an end-to-end, extensible, coherent security framework that enables the protection of the platform at all times during its life cycle. Such a framework must start during manufacturing, where it is verified that the data loaded onto the platform is both authentic and authorized. During provisioning and use of a trusted mobile platform, the platform measures the integrity of its hardware as well software to ensure that it is not corrupted by malware, enforces safe handling of valuable private data, processes secrets in a protected environment to ensure that keys and data are not observed, corrupted, or stolen. Such a platform also has a reliable physical protection that ensures that its hardware- and software-based security mechanisms cannot be bypassed or disabled.

A trusted mobile platform also has verifiable software and hardware configuration, behaves in a specific and predictable way, and guarantees the correctness, accuracy, and authenticity of the security and system components. With these attributes, such a platform could validate its own correctness as well as the correctness of its security building blocks. How can we implement such a platform? Let's study some options.

Adding a hardware security module to the platform can be inflexible and can have an adverse effect on power consumption. Also, with a hardware-only solution, if an error is discovered, it cannot be easily fixed.

Adding off-chip hardware, such as cryptographic coprocessors and storage, can add to the cost, complexity, and power budget of the platform. Furthermore, the traffic between the core processor and the off-chip device can be exposed. In such situations, it is not possible to confirm the integrity of the off-chip device, which can be removed and interfered with.

Several researchers have explored the idea of using hardware to ensure platform trustworthiness. Herzberg and Pinter [5] describe a device that can be used to protect software against piracy. Chaum and Pedersen [6] describe a trusted, hardware-based wallet architecture that both carries a database with personal information and protects the database from unauthorized access. Yee and Tygar's approach [7] ensures that the system functions securely. Sander and Tschudin [8] describe a code-protection mechanism that relies on the execution of encrypted functions and does not need trusted hardware. The approach proposed by Hohl [9] allows arbitrarily complex functions but guarantees protection only for a certain time interval. Cahill et al. [10] describe a system for dynamically assessing risk and trustworthiness based on various types of evidence, some of which is assumed to be gathered from previous experience.

Trust must be an end-to-end, integral part of the mobile platform [11–13] and therefore cannot be treated as a separate add-on hardware or software. For example, on the communication aspect of the platform trust, the integrity of the data exchange must be guaranteed, and participants exchanging data must be authenticated and their actions must be authorized.

Applications running on a trusted platform must be constrained so that they can only access the platform capabilities that they need and they are restricted to the specific space where they are running.

The following sections describe some emerging mobile applications that require some level of trust in the mobile platforms, networks, and protocols. More details about these applications are available in the follow-up chapters.

1.2.3 M-Commerce Applications

New mobile payment technologies address retail payment transactions, such as micropayments. These technologies offer the ability to pay for products, vending, ticketing, mobile content services, and games or gambling. The service charges are typically billed on a mobile phone bill or on a credit card bill. An alternative approach is for the customer to open a separate customer account, commonly named mobile wallet, which allows the user to transfer money from a bank account.

M-commerce can cover all main financial payment methods such as cash, direct debit, credit cards, and payment of service bill. Consumers, financial institutions, merchants, and service providers can benefit from such mobile payment solutions. Therefore, problems with payments can severely hamper the deployment of M-commerce.

The security risks for such financial transactions performed on a mobile platform can include unauthorized use, transaction errors, lack of transaction records and documentation, privacy issues, and device and mobile network reliability. Chapters 4 and 5 provide more insight into the protection mechanisms available for M-commerce.

1.2.4   Mobile Web Services Applications

Web services are loosely coupled computing services that can reduce the complexity of building business applications and enabling new business models. In Web services, the interaction between distant applications is instantaneous, since interaction is more from application to application rather than from humans to applications.

In contrast to client-server server communications, extensible markup language (XML)–based Web services are designed to seamlessly connect resources above the network layer, thus enabling the concept of loosely coupled but tightly contracted applications. These applications enable easy direct access to backend databases and application servers.

Securing Web services is crucial for financial, legislative, privacy, and trust reasons. Existing security standards, such as the secure sockets layer (SSL), are not adequate for securing Web services because Web services require end-to-end security rather than the point-to-point security provided by SSL. In the Web services context, security means that a message recipient is able to successfully and securely perform the following operations: verify the integrity of a message, receive a message confidentially, determine the identity of the sender, and determine if the sender is authorized to perform the operation requested by the message. Chapter 5 provides more details on how to protect mobile Web services applications.

1.2.5   Mobile Digital Rights Management Applications

The sharing of media and entertainment via mobile platforms is becoming an increasingly popular mobile service. Users of mobile platforms download content to their mobile phones or receive information by using emerging services such as multimedia messaging subsystem (MMS).

Wireless carriers and content providers lose revenue when premium content is forwarded from one user to the next for free. These carriers and content providers need to ensure the protection of their intellectual property and provide integrity protection to that content while in transit or when residing on a customer's mobile platform. Their customers also require security from malicious acts as well as privacy.

Digital rights management (DRM) solutions must provide those protections, as well as work across different devices, geographies, operators, and mobile platforms. They need to protect content wherever it travels and enforce administrator-defined policies. Without a secure and interoperable DRM solution, the full potential of mobile media and entertainment delivery cannot be realized. Chapter 5 provides more details on mobile DRM.

1.3 Overview of the Book

This book has been written to appeal to a wide audience. Different chapters of the book are targeted to different types of readers and can be read in conjunction with the other chapters or alone.

There are sixteen chapters in this book. Chapter 1 is an introduction to the book. Chapters 2 through 14 delve into the technologies necessary to understand mobile and wireless security. Chapter 2 explains the basics of identification, authentication, authorization, and non-repudiation. Chapter 3 provides a good introduction to the cryptographic techniques commonly used in mobile protocols and devices. Chapters 4 and 5 explain hardware security and software security, respectively. Chapter 6 provides some background on security certification, which is increasingly gaining importance for security systems.

Chapters 7 and 8 explain higher layer security and IP layer security. Chapter 9 describes authentication–authorization from networking and protocol perspectives. Chapter 10 provides a detailed description of EAP and 802.1X. Chapters 11 through 14 focus on the security aspects of wireless area networks. Chapter 11 covers WPAN security, such as Bluetooth, Zigbee, and ultrawideband. Chapter 12 describes WLAN security, particularly 802.11. Chapter 13 provides an exposition of the end-to-end security aspects of WMANs, such as WiMAX. Chapter 14 covers WWAN security, which includes GSM and 3G technologies.

In Chapter 15, the authors provide their perspective and predictions on future wireless technologies and related security challenges. Finally, Chapter 16 provides a unique mobile security threat catalog, which is a list with descriptions of known threats for mobile platforms and communication. This catalog also includes references to the book's chapters where the topic is addressed.

References

[1]   Spafford, G., http://homes.cerias.purdue.edu/~spaf/quotes.html.

[2]   ISO/IEC 2382-8. http://www.iso.org.

[3]   RFC 2828. http://www.faqs.org/rfcs/rfc2828.html.

[4]   U.S. Secret Service Financial Crimes Division, http://www.secretservice.gov/financial_crimes.shtml#Telecommunications.

[5]   Herzberg, A., and S. S. Pinter, "Public Protection of Software: Advances in Cryptology—Crypto 85," *Lecture Notes in Computer Science,* Vol. 218, 1985.

[6]   Chaum, D., and T. P. Pedersen, "Wallet Databases with Observers: Advances in Cryptology—Crypto 92," *Lecture Notes in Computer Science,* Vol. 740, 1992.

[7]   Yee, B., and D. Tygar. "Secure Coprocessors in Electronic Commerce Applications," *Proceedings of the First USENIX Workshop on Electronic Commerce,* USENIX Association, Berkeley, California, 1995.

[8]   Sander, T., and C. Tschudin, "Toward Mobile Cryptography," *IEEE Symposium on Security and Privacy,* IEEE Computer Society, Los Alamitos, CA, 1998.

[9]   Hohl, F., "Time Limited Blackbox Security: Protecting Mobile Agents from Malicious Hosts, Mobile Agents and Security," Vol. 1419, *Lecture Notes in Computer Science,* G. Vigna (ed.), Berlin: Springer-Verlag, 1998.

[10] Cahill, V., et al., "Using Trust for Secure Collaboration in Uncertain Environments," *IEEE Pervasive Computing,* Vol. 2, No. 3, July–September 2003.

[11] Aissi, S., et al., "Trusted Mobile Platform Protocol Specification," http://xml.coverpages.org/TMP-ProtocolV10.pdf.

[12] Aissi, S., et al., "Trusted Mobile Platform Software Architecture Description," http://www.trusted-mobile.org/TMP_SWAD_rev1_00.pdf.

[13] Aissi, S., et al. "Trusted Mobile Platform Hardware Architecture Description," http://xml.coverpages.org/TMP-HWADv10.pdf.

2

Authentication, Authorization, and Non-Repudiation

2.1 Introduction

Authentication is the process of verifying a user's identity, which typically entails obtaining a user name and a password or some other credential from the user. Authorization is the process of verifying whether that user has access to some protected resources. Non-repudiation is the process of ensuring that the sender of a transaction is not able to falsely deny later that he sent the transaction and that the receiver is not able to falsely deny later that she didn't receive the transaction.

Authentication is based on each user having a unique set of criteria for gaining access to a device, network, or service. Once the user provides his credentials, they are compared with other user credentials stored in a database. If the credentials match, the user is granted access; otherwise, authentication fails and access is denied.

After authentication, the user must gain authorization for accessing some services or performing certain tasks. The authorization process determines whether he has the authority to perform such tasks or to access such services as commands. Hence, authorization is the process of enforcing policies by determining what types or of resources, activities, or services the user is permitted.

2.2 Authentication

Authentication is necessary because it allows verification of the identity of an entity, user, or device before access to resources and services can be granted. As shown in Figure 2.1, two entities are involved in the authentication process: the supplicant is the one who wants to prove her identity and the authenticator is the one that needs to verify it. The authenticator can be local or remote.

Authentication can be one way or mutual, depending on whether only one or both entities involved in the communication prove their identity. Authentication can be performed by multiple means, the most simple of which consists of using a password, where it is expected that only a single entity knows the password linked it its identity. To keep a password secret, it must be transmitted over an encrypted channel, but other ways have been found to avoid the creation of privacy-protected communication for authentication. One-time passwords (OTPs), whose lifetimes are very short, can be transmitted in clear.

The peers involved can perform a challenge-response algorithm to demonstrate their identities. The algorithm basically consists of a response to a challenge sent from the supplicant to authenticator. The correct response can only be calculated by the entity who knows the key associated with the peer's identity. As with the use of an OTP, a challenge-response algorithm can be performed over nonsecure means.

EAP methods allow the performance of authentication according to EAP format. EAP was standardized by the Internet Engineering Task Force (IETF), whereby a general framework and packet format was defined to allow great flexibility in the acceptance of the authentication algorithms. The algorithm choice depends on the EAP method employed; tens have been defined, some of which are standard while others are proprietary.

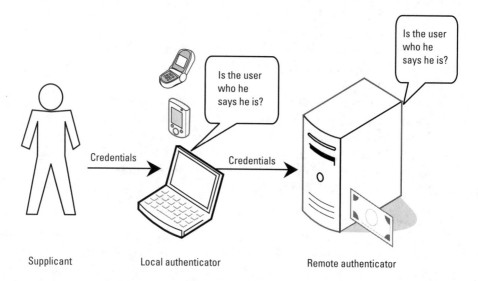

Figure 2.1 Authentication.

In the following section, we will describe password and challenge-response authentication, whereas EAP methods will be discussed in a dedicated chapter.

2.2.1 Static and One-Time Passwords

The easiest authentication means consists of the use of passwords: an entity proves it is who it claims to be by transmitting the password associated with its identity. The drawbacks are twofold: password management and their confidentiality.

Before password-based authentication can be deployed, a distribution procedure must be in place. Passwords must be transmitted to the associated entities in a secure way. The appropriate means to use depends strongly on the application scenario. When used for home banking, for example, passwords can be sent through the mail. In the enterprise world, they are often communicated over the phone to users or directly programmed on the machines by computer system employees. Although basic mail and phone systems do not rely on security services such as encryption or authentication, they are practical and are perceived as reliable because an attacker must physically intercept the communication to retrieve the information. If a secure electronic channel already exists between the password management unit and the entities involved, the password can be transmitted over this channel.

The privacy of static passwords must be protected throughout their lifetime. This involves protecting the password while stored and when transmitted. Passwords should be memorized by users or recorded on protected storage units. When transmitted for authentication, they must be sent over an encrypted channel to the authenticator. If the password is eavesdropped, a mechanism must be in place to update it and avoid the use of the broken password.

Passwords are vulnerable to dictionary attacks: since most often users pick their own passwords they tend to choose easy to remember strings, such as common words or dates. To avoid dictionary attacks, real random alphanumeric sequences should be chosen.

Personal identification numbers (PINs) are a subclass of passwords where only digits are used. PINs are often used so that a user can identify himself as the possessor of a portable device such as a mobile phone or banking card. In these cases, PINs are used to reduce theft and to forbid malevolent users access to the resources.

OTPs are a possible alternative to static passwords. The advantage is that their confidentiality does not have to be guaranteed because of their very short lifetime. The lifetime of an OTP can be equivalent to a certain timeframe, usually minutes or seconds, or OTPs can be valid once, independently of the time they are employed. In both cases, synchronization between the supplicant and the authenticator is necessary.

OTPs are usually generated as the output of an algorithm whose inputs are a seed, frequently the clock or a counter, and a master key. The algorithm can be programmed in software, or a dedicated hardware device can be deployed for this purpose.

Password distribution remains an issue when using OTPs for authentication. Usually computer system employees configure the machines with the user's respective master keys, or they distribute hardware tokens. OTP's main use is user verification to access an enterprise private network. Commercial solutions on the market include the SecurID authentication token by RSA as well as the VeriSign OTP Token.

2.2.2 Challenge-Response Authentication

In challenge-response authentication, the authenticator sends a random challenge to the supplicant, who authenticates himself by returning the response on the received challenge. Both peers must know which key and which algorithm must be used to calculate the response, and the key is unknown to all other parties except for the involved ones [1].

The secret key must be distributed to the parties before authentication can take place. The key could have been previously distributed by any means; it could be configured into the authentication program, it could be the result of a Diffie-Hellman (DH) key agreement, or it could have been transmitted encrypted using public key encryption.

Different algorithms could be used to calculate the challenge; most often secret key encryption algorithms or HMACs are preferred. The authentication security will be related to both the algorithm strength and key length used.

The challenge and the response can be transmitted clearly; if a secure algorithm was used, their value will not reveal any information on the authentication key. Care must be taken to use fresh random values for authentication, or the risk of replay attacks is encountered. There must also be a good random number generator at the authenticator side to reduce the chances of analysis on previously transmitted responses to obtain information on the secret key.

If mutual authentication should be performed, the challenge response can be calculated twice, switching supplicant and authenticator roles.

Challenge-response authentication algorithms are often used in wireless technologies. In the following chapters we will see that GSM requests client authentication and that 3GPP requests mutual authentication using challenge-response schemes. Bluetooth authentication also relies on challenge response. Historically, 802.11 and 802.16 relied on this scheme as well, but because of the complexity of key distribution, the most recent enhancements suggest use of EAP methods.

2.3 Authorization

2.3.1 Introduction

Authorization is often called access control. Following authentication, authorization is used to determine what the identified user has access to and with what privileges or authorities. Authorization is the process of determining if a particular right, such as access to some resource, can be granted to a supplicant who has presented a particular credential. Logically, authorization is preceded by authentication [2].

As shown in Figure 2.2, two entities are involved in the authorization process: the supplicant who wants to access some resources or perform a task, and the authorizer who can give it access. The authorizer can be local or remote, and authorization to access some resources depends on the supplicant's account rights and privileges.

2.3.2 Authorization Strategies

In the enterprise environment, a system administrator defines (based on corporate policies) which users are allowed access to any particular system and what privileges they may use, such as access to which directories, access times, storage allocation, and execution rights. On the user's mobile platform, an application identifies what resources the user can be given during this session based on the permissions that were provided to her. Therefore, authorization is both the

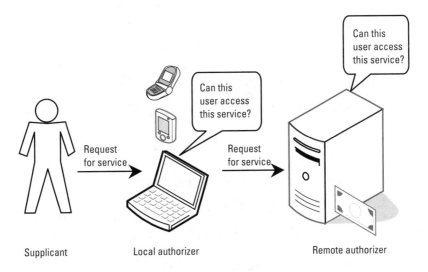

Figure 2.2 Authorization.

preliminary configuration of permissions by a system administrator and the actual checking of the permissions when the user is gaining access.

Authorization requires strong identification of the resources that need to be exposed to a supplicant. Resources can include server resources (e.g., user or application data), services (e.g., online banking), and static resources (e.g., remote files).

There are several authorization strategies, but role-based authorization is the most common one. In this authorization strategy, users are mapped into roles. Access to resources is secured based on the role membership of the supplicant. Roles are used to partition the resource's user base into sets of users that share the same security privileges within the resource. This helps the strategy's scalability tremendously, since authorization for access is based on role membership rather than directly tied to specific resources.

2.3.3   AAA Authorization

The authorization model developed by the Authentication, Authorization, and Accounting (AAA) working group [3–5] is a centralized one. In this model, there are four entities:

- The user who needs to be authorized to access a service;

- The user's home organization with which the user has an agreement;

- The AAA server of the service provider, which authorizes the service based on an agreement with the user's home organization;

- The service equipment of the service provider, which provides the service itself.

The main authorization requirements for an AAA server are supporting combined authorization and authentication messages, allowing to forward requests to another AAA server, and allowing intermediate brokers to add their own authorization information to the requests and to the replies.

AAA has three different kinds of authorization sequences: agent, pull, and push.

In the agent sequence, the AAA server acts as an agent between the user and the service equipment. In the pull sequence, the user sends a request to the service equipment, which forwards it to the AAA server. The AAA server decides whether or not to allow the service and sends the response to the service equipment. The service equipment sets up the service, if allowed, and alerts the user that the service is ready for use. In the push sequence, the user gets a ticket from the AAA server indicating that that he is permitted to access the service. Then, the user sends the request as well as the ticket to the service equipment, which verifies the ticket. If the ticket verification is successful, the service equipment sets up the service and sends a reply to the user indicating that the service is available for use.

2.4 Non-Repudiation

2.4.1 Introduction

Because of the potential ubiquity of mobile and wireless platforms, users could send more messages, more often, and more easily. Furthermore, M-commerce transactions will be more prevalent than paper-based transactions.

Message authentication, integrity, and non-repudiation are required for such transactions. Non-repudiation (also called non-repudiability) is a means to ensure that the sender of a transaction is not able to falsely deny later that she sent the transaction and that the receiver is not able to falsely deny later that he didn't recive the transaction. By definition, in reference to digital security, non-repudiation means to ensure that a transferred message has been sent and received by the parties claiming to have sent and received the message [6].

From a technical perspective, non-repudiation of submission and non-repudiation of receipt can be achieved by tightly binding transactions and customers, making transactions difficult to forge, making transactions unalterable, and making transactions verifiable.

Non-repudiation is a well-defined concept in information security; however, it has never been defined as a core business requirement for financial transactions in many countries. The main reasons are that repudiation disputes are commonly resolved by legal and contractual agreements, rather than technical evidence, and that implementing non-repudiation for low-value financial transactions is usually not very cost effective. Furthermore, there's an obvious lack of global legislation that requires a specific technology for non-repudiation and a clear lack of a technical standard that defines such technology for legal binding.

This section describes how non-repudiation can be achieved and therefore how to guarantee that the sender of a message cannot later deny having sent the message (non-repudiation of origin) and that the recipient cannot deny having received the message (non-repudiation of destination).

2.4.2 Non-Repudiation Methods

The notion of non-repudiation for an electronic transaction is to prevent two entities, a customer and a service provider, from repudiating the transaction after it is committed. Typically, that can be achieved through the use of digital signatures, confirmation services, timestamps, and audits.

Digital signatures work as a unique identifier for an individual, much like a written signature. Confirmation services can be achieved when the message transfer agent or protocol creates digital receipts to indicate that messages were sent or received. Timestamps contain the date and time a document was composed and prove that a document existed at a certain time. The auditing component of non-repudiation provides the essential capability of collecting, recording, analyzing, and responding to all reported events during a transaction or message exchange.

2.4.3   Digital Signatures

A common way to implement non-repudiation is to use digital signatures. In a way, digital signatures could be considered a replacement to the traditional signature in electronic data processing. Using digital signatures, when a customer sends a service provider a signed document, the service provider knows that a specific customer is the one who sent it, because the signature contains his public key. Also, since the signature is based on the contents of the document, the signature will not match if any changes have been made to the document.

Typically, digital signatures are enabled using a trusted third party (TTP) or public key infrastructure (PKI), which usually support at a minimum a certification authority (CA) for issuing the digital certificates as well as the certificate revocation lists (CRLs). Digital certificates provide strong binding between the device owner or operator and the public keys. On mobile platforms, the private key(s) corresponding to those public keys must be securely stored. Smart cards and trusted platform modules (TPMs), for example, can be used to safeguard those keys. In the absence of such trusted hardware, proper care should be used in implementing a software solution to protect the signing private keys [7].

Special care needs to be considered when implementing a digital signature solution for high-value financial services on mobile platforms. Digital signatures may be vulnerable to forgery and are also potentially subject to fraud. On a mobile platform that has been broken into or infected with malware, digital signatures may be forged.

2.4.4   Timestamps

The ability to prove the date and time on which a transaction has been made, or that a specific message has been sent or received, is of the utmost importance for non-repudiation. Timestamps are digitally signed files that vouch for the existence of a specific transaction at a specified date and time. All parties in such a transaction must be able to prove the time and date a transaction took place.

Timestamping provides non-repudiation services by attesting to the existence of a transaction at a particular time. Timestamping typically achieves this by cryptographically linking an imprint of a document along with a timestamp.

2.4.5 Auditing

Auditing is the comprehensive secure logging of transaction events. It must track, record, and report on all transactions. Furthermore, it must have the ability to scrutinize those events. The auditing function should also capture warnings and errors in addition to positive events such as successful transactions.

The audit records of all such events must be kept in a secure data store and must include tools for accessing and leveraging the logged data. The integrity of such data must be ensured, for example, by using digital signatures. Typical auditing services also provide real-time event information for notification and monitoring. Such trigger notifications may be based on predefined event filtering.

2.4.6 Related Standards

IETF RFC 3161 [8] defines a time stamp authority (TSA) that can be used to associate a transaction with a particular point in time. This TTP provides a proof of existence for this particular transaction at an instant in time.

IETF RFC 2459 [9] specifies two bits that are critical for non-repudiation within the KeyUsage extension: the *nonRepudiation* bit and *digitalSignature* bit. The *nonRepudiation* bit is asserted when the subject public key is used to verify digital signatures that are used to provide a non-repudiation service that protects against the signing entity falsely denying some action, excluding certificate or CRL signing. The *digitalSignature* bit is asserted when the subject public key is used with a digital signature mechanism to support security services other than non-repudiation, certificate signing, or revocation information signing.

ISO/IEC 13888 [10] provides a general model for specifying non-repudiation mechanisms using cryptographic techniques. In this standard, the purpose of non-repudiation is to provide verifiable proof or evidence recording of data, based on cryptographic check values generated using cryptographic techniques.

References

[1] Menezes, A., P. van Oorschot, and S. Vanstone, *Handbook of Applied Cryptography,* Boca Raton, FL: CRC Press, 1997.

[2] Norton, P., and M. Stockman, *Network Security Fundamentals,* 1st Ed., Indianapolis, IN: SAMS, 2000.

[3] IETF RFC 2903, Generic AAA Architecture, August 2000.

[4]   IETF RFC 2904, AAA Authorization Framework, August 2000.

[5]   IETF RFC 2905, AAA Authorization Application Examples, August 2000.

[6]   Online dictionary for computer and Internet technology: http://www.webopedia.com.

[7]   Tipton, H., and M. Krause, *Information Security Management Handbook,* 4th Ed., Boac Raton FL, Auerbach Publications.

[8]   IETF RFC 3161, Internet X.509 Public Key Infrastructure—Time-Stamp Protocol (TSP).

[9]   IETF RFC 2459, Internet X.509 Public Key Infrastructure—Certificate and CRL Profile (CCP).

[10] ISO/IEC 13888-1-2-3, Information Technology—Security Techniques—Non-Repudiation.

3

Cryptographic Techniques

Mobile communication users expect the same security as that granted for fixed communications. On dedicated fixed lines, tampering is possible by eavesdropping on the physical line. Wireless communications are more vulnerable to attacks because they can be performed remotely and over the air. By attacking a wireless technology once, information on the communications of multiple users worldwide can be obtained. A much greater effort is needed to breach as many fixed communication lines.

Even more than in the past, today information is one of the world's greatest assets and must therefore be protected. Historically, the first service identified when dealing with security is confidentiality, obtained by encryption. All the mobile communication systems described in this chapter provide means to conceal the user information transmitted. Nevertheless, user information is not the only data exchanged within a communication. Control and management data to set up and monitor the communication channel are also sent but are rarely secured. Although this lack does not endanger the secrecy of user data, it may allow performance of attacks such as DoS. Encryption does not guarantee that data transmitted will not be corrupted by active attacks or erroneous transmissions. Integrity protection techniques should be implemented to allow the receiver to verify whether the data that reached her matches what was sent. Last but not least, the parties involved in a communication must have a way to identify each other. Voice tone, a unique characteristic for each person, can be used in person-to-person exchanges, but authentication systems must be relied on in all other cases. Authentication is not only a service offered to the user, but a necessary mechanism for the service provider's billing system.

The security services expected from a mobile communication standard are authentication, encryption, and integrity. As a matter of fact, some of the technologies we describe in this book, such as GSM and 802.11, were defined in the nineties, when security was not considered a primary issue in civil communications. A decade ago, available resources were mainly dedicated to performance optimization, and consequently security, which inevitably takes away precious cycles, was minimized. Security is also often reduced due to compatibility needs within the same technology (i.e., not all services may be available, and the strength of the services offered may be limited).

In this chapter we describe algorithms employed in wireless technologies, and we evaluate their security. The security protocols in which the algorithms are employed are not described herein but can be found in each technology's specific chapter (i.e., Chapter 11 for Bluetooth, Chapter 12 for Wi-Fi security, and Chapter 14 for GSM- and 3GPP-related algorithms).

3.1 GSM Algorithms

Global system for mobile communications (GSM) security provides user-identity authentication and voice encryption. It is assumed that the network is secure on the mobile-operator side, as no network security is envisaged in the standard.

GSM security is based on a 128-bit secret key, referred to as *Ki,* shared by the network operator and the mobile user. *Ki* is the secret key used in the challenge response algorithm to authenticate the mobile user and calculate the voice encryption key. *Ki*'s confidentiality is crucial, as GSM security depends uniquely on it. The network must be secure on the mobile operator side. The subscriber identity module (SIM) is the tamper-resistant device on the user side that stores sensitive data and performs cryptographic operations.

Three algorithms are defined for GSM security and are described in [1]:

- A3 is the authentication algorithm;
- A5 is the ciphering/deciphering algorithm;
- A8 is the ciphering key generator.

A unique algorithm, referred to as A3/A8, is often used both for authentication and ciphering key generation.

Algorithms A3 and A8 can be implemented at the operator's discretion; the specifications [1] only define their input and output format. The GSM association has nevertheless defined possible A3/A8 implementations, namely the COMP 128 series, available to GSM network operators and manufacturers of GSM equipment.

Algorithm A5 must be common to all GSM public land mobile network (PLMN) stations and mobile stations to allow roaming. Early A5 implementation specifications (i.e., versions 1 and 2) are available for GSM association members, whereas version 3 is publicly available [2]. The A5/3 algorithm is based on the block cipher KASUMI [3].

GSM authentication protocol and encryption key generation are described in Chapter 14. Herein we will describe the A5 algorithm in its latest version, version 3.

The GSM A5/3 algorithm produces two 114-bit keystream strings, one used for uplink data and the other for downlink data. The input to the keystream generator is the GSM encryption key K_c, which has to be previously generated by the algorithm A8.

3.1.1 The A5/3 Algorithm

The A5/3 algorithm is a stream cipher;[1] encryption and decryption occur by xoring the plaintext or ciphertext with a keystrem. The keystream is generated by applying the algorithm represented in Figure 3.1, where *INIT* are initialization and frame dependent counter bits, *KM* is a constant key modifier value,

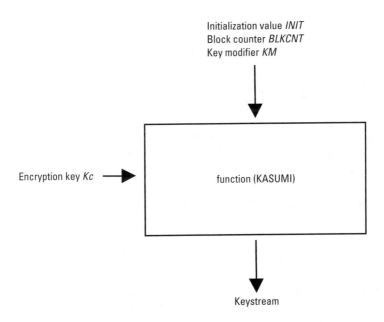

Figure 3.1 The A5/3 keystream generator.

---

1. For stream cipher definition refer to [4].

BLKCNT is the block counter value from 0 to 3 and K_c is the 64-bit encryption key generated by the A8 algorithm.

3.1.2   KASUMI

KASUMI [3] is a block cipher developed by the 3GPP Task Force for confidentiality and integrity protection. It is a Feistel cipher[2] with eight rounds, operating on a 64-bit data block using a 128-bit key.

KASUMI is decomposed in three subfunctions *FL, FO, FI* used with subkeys *KL, KO, KI.*

The input *I* is divided into two 32-bit strings L_0 and R_0, where $I = L_0 \| R_0$.
Then for each integer i with $1 \leq i \leq 8$ we define

$$R_i = L_{i-1} \tag{3.1}$$

$$L_i = R_{i-1} \oplus f_i\left(L_{i-1}, RK_i\right) \tag{3.2}$$

This constitutes the ith round function of KASUMI, where f_i denotes the round function with L_{i-1} and round key RK_i as inputs.

The result *OUTPUT* is equal to the 64-bit string $(L_8 \| R_8)$ obtained at the end of the eighth round.

The structure of KASUMI is reported in Figure 3.2.

3.1.2.1   Key Schedule

Each round of KASUMI uses a 128-bit subkey RK_i, derived from the 128-key K, where $RK_i = KL_i, KO_i, KI_i$.

The bit-by-bit operations necessary to calculate the subkeys are shown in Table 3.1.

The values *K1…K8* are subdivisions of the key *K;* that is,

$$K = K1\|K2\|K3\|...\|K8 \tag{3.3}$$

The values *Kj′* are derived by applying the following:
For each integer j with $1 \leq j \leq 8$

$$Kj' = Kj \oplus Cj \tag{3.4}$$

where *Cj* is a constant value.

---

2. For Feistel cipher definition, refer to [4].

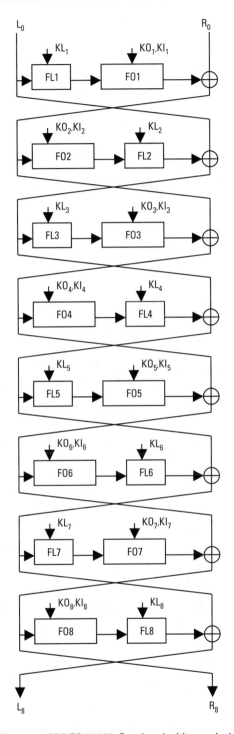

Figure 3.2 KASUMI. (Source: 3GPP TS 35.202. Reprinted with permission from 3GPP.)

Table 3.1
Round Subkeys

	1	2	3	4	5	6	7	8
$KL_{i,1}$	$K1<<<1$	$K2<<<1$	$K3<<<1$	$K4<<<1$	$K5<<<1$	$K6<<<1$	$K7<<<1$	$K8<<<1$
$KL_{i,2}$	$K3'$	$K4'$	$K5'$	$K6'$	$K7'$	$K8'$	$K1'$	$K2'$
$KO_{i,1}$	$K2<<<5$	$K3<<<5$	$K4<<<5$	$K5<<<5$	$K6<<<5$	$K7<<<5$	$K8<<<5$	$K1<<<5$
$KO_{i,2}$	$K6<<<8$	$K7<<<8$	$K8<<<8$	$K1<<<8$	$K2<<<8$	$K3<<<8$	$K4<<<8$	$K5<<<$
$KO_{i,3}$	$K7<<<13$	$K8<<<13$	$K1<<<13$	$K2<<<13$	$K3<<<13$	$K4<<<3$	$K5<<<13$	$K6<<<13$
$KI_{i,1}$	$K5'$	$K6'$	$K7'$	$K8'$	$K1'$	$K2'$	$K3'$	$K4'$
$KI_{i,2}$	$K4'$	$K5'$	$K6'$	$K7'$	$K8'$	$K1'$	$K2'$	$K3'$
$KI_{i,3}$	$K8'$	$K1'$	$K2'$	$K3'$	$K4'$	$K5'$	$K6'$	$K7'$

From 3GPP TS 35.202

3.1.2.2   Function f_i

The function $f_i(\ )$ takes a 32-bit input I and returns a 32-bit output O under the control of a round key RK_i. The function itself is constructed from two subfunctions: FL and FO. The subkey KL_i is used with FL and the subkeys KO_i and KI_i are used with FO.

The $f_i(\ )$ function has two different forms, depending on whether it is an even round or an odd round.

For odd rounds the data is passed through $FL(\ )$ and then $FO(\ )$; that is, for $i = 1, 3, 5, 7$:

$$f_i(I, RK_i) = FO\left(FL(I, KL_i), KO_i, KI_i\right) \tag{3.5}$$

and for even rounds the data is passed through $FO(\ )$ and then $FL(\ )$; that is, for $i = 2, 4, 6, 8$:

$$f_i(I, K_i) = FL\left(FO(I, KO_i, KI_i), KI_i\right) \tag{3.6}$$

3.1.2.3   Function FL

The input to the function FL comprises a 32-bit data input I and a 32-bit subkey KL_i. The subkey KL_i and the input data I are split into two 16-bit halves, $KL_i = KL_{i,1} \| KL_{i,2}$ and $I = L \| R$.

We define

$$R' = R \oplus ROL\left(L \cap KL_{i,1}\right) \tag{3.7}$$

$$L' = L \oplus ROL\left(R' \cup KL_{i,2}\right) \tag{3.8}$$

The 32-bit output value is $(L'\|R')$.

3.1.2.4 Function *FO*

The input to the function *FO* comprises a 32-bit data input I and two sets of subkeys: a 48-bit subkey KO_i and 48-bit subkey KI_i.

The input data and subkeys are subdivided into 16-bit halves; that is,

$$I = L_0\|R_0, KO_i = KO_{i,1}\|KO_{i,2}\|KO_{i,3} \text{ and } KI_i = KI_{i,1}\|KI_{i,2}\|KI_{i,3}.$$

Then for each integer j with $1 \le j \le 3$, we define

$$R_j = FI\left(L_{j-1} \oplus KO_{ij}, KI_{i,j}\right) \oplus R_{j-1} \tag{3.9}$$

$$L_j = R_{j-1} \tag{3.10}$$

Finally, we return the 32-bit value $(L_3\|R_3)$.

3.1.2.5 Function *FI*

The function *FI* takes a 16-bit data input I and 16-bit subkey $KI_{i,j}$. The input I is split into two unequal components, a 9-bit left half L_0 and a 7-bit right half R_0, where $I = L_0\|R_0$.

Similarly the key $KI_{i,j}$ is split into a 7-bit component $KI_{i,j,1}$ and a 9-bit component $KI_{i,j,2}$, where $KI_{ij} = KI_{ij,1}\|KI_{ij,2}$.

The function uses two *S*-boxes: *S*7, which maps a 7-bit input to a 7-bit output, and *S*9, which maps a 9-bit input to a 9-bit output. The function *FI* also uses two additional functions, which we designate *ZE*() and *TR*(). We define these as follows:

- *ZE*(x) takes the 7-bit value x and converts it to a 9-bit value by adding two zero bits to the most-significant end.
- *TR*(x) takes the 9-bit value x and converts it to a 7-bit value by discarding the two most-significant bits.

We define the following series of operations:

$$L_1 = R_0 \tag{3.11}$$

$$R_1 = S9[L_0] \oplus ZE(R_0) \tag{3.12}$$

$$L_2 = R_1 \oplus KI_{ij,2} \tag{3.13}$$

$$R_2 = S7[L_1] \oplus TR(R_1)KI_{ij,1} \qquad (3.14)$$

$$L_3 = R_2 \qquad (3.15)$$

$$R_3 = S9[L_2]ZE(R_2) \qquad (3.16)$$

$$L_4 = S7[L_3] \oplus TR(R_3) \qquad (3.17)$$

$$R_4 = R_3 \qquad (3.18)$$

The function returns the 16-bit value $(L_4 \| R_4)$.

3.2 3 GPP Algorithms

The Third Generation Partnership Project (3GPP) security provides user and network mutual authentication, voice and data encryption, and integrity protection. 3GPP Authentication and Key Agreement (AKA) is based on a 128-bit secret key K shared between the user and the operator. Given a random value RAND, the quintet $Q = (RAND, XRES, CK, IK, AUTN)$ is calculated by applying security functions using the key K. Specifically, *RAND* is the network-chosen random challenge, *XRES* is the expected user response, *CK* the encryption key, *IK* the integrity key, and *AUTN* the network authentication token.

3GPP AKA will be examined in detail in Chapter 14, where we will deal with the security protocol and services provided as a result of it. The goal of this section is to describe the MILENAGE algorithm, that is, the 3GPP security algorithm. 3GPP security is based on a computation of quintet Q, for which the 3GPP specifications [5] have defined seven security functions $f1, f1^*, f2, f3, f4, f5,$ and $f5^*$. AUTN is not a direct output of MILENAGE, but it is calculated as linear combination of MILENAGE outputs.

$$AUTN = (SQN \oplus AK)\|AMF\|MAC - A \qquad (3.19)$$

As authentication concerns only the operator providing the service, each operator is free to choose separately the function implementation. 3GPP has nevertheless defined example functions that can be adopted by operators who do not wish to design their own.

3.2.1 MILENAGE

The MILENAGE algorithm defined by 3GPP makes use of:

- A block cipher encryption function E, which takes a 128-bit input x and a 128-bit key k and returns a 128-bit output $E[x]_k$. The block cipher selected is Advanced Encryption System (AES) [6] with 128-bit key and 128-bit block size. If desired, an operator may use an algorithm of his choice other than AES.

- A 128-bit value operator variant algorithm configuration field open platform (OP). It is up to operators to decide how to manage OP: it can be a secret or publicly known value, fixed or variable, constant or different for each user.

- Five 128-bit constants $c1$, $c2$, $c3$, $c4$, $c5$ and five integers $r1$, $r2$, $r3$, $r4$, $r5$. Reference values are defined in the MILENAGE specification [5], but, again, operators may choose other values.

- A 48-bit fresh sequence number SQN generated by the operator authentication center.

- A 16-bit authentication management field AMF set by the operator. The use of AMF is not standardized and can be defined separately by each operator.

Function by function, MILENAGE's output is

- Output of $f1 = MAC\text{-}A$, the 64-bit message authentication code used for authentication of the network to the user.

- Output of $f1^* = MAC\text{-}S$, the 64-bit message authentication code used to provide data integrity and data origin authentication for verification of the user by the authentication center.

- Output of $f2 = RES$, the user response. Its length is a multiple of 8 bits, at least 32 bits and at the most 128 bits.

- Output of $f3 = CK$, the 128-bit cipher key.

- Output of $f4 = IK$, the 128-bit integrity key.

- Output of $f5 = AK$, the 48-bit anonymity key.

- Output of $f5^* = AK$, the 48-bit anonymity key.

AK is the name of the output of either $f5$ or $f5^*$; in practice these two functions will not be computed simultaneously.

In Figure 3.3, the input value OPC is calculated as $OP_C = OP \oplus E[OP]_K$.

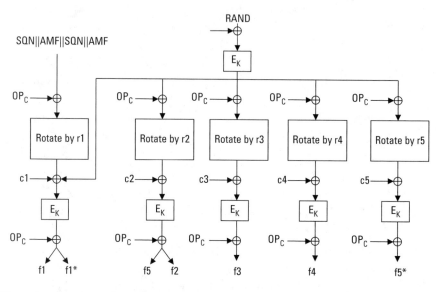

Figure 3.3 MILENAGE block diagram. (Source: 3GPP TS 35.206. Reprinted with permission from 3GPP.)

3.2.2 GPP Encryption and Integrity Functions

MILENAGE allows the operator and user to share an encryption key CK, which will be used with the confidentiality algorithm $f8$, and an integrity key IK, which will be used with the integrity algorithm $f9$. For compatibility between operators for roaming purposes, these algorithms have been fully standardized [7]. Each of these algorithms is based on the KASUMI algorithm described earlier.

The confidentiality algorithm $f8$ is a stream cipher that is used to encrypt/decrypt blocks of data between 1 and 20,000 bits long under the confidentiality key CK. KASUMI is used in a form of output-feedback mode as a keystream generator.

The integrity algorithm $f9$ computes a 32-bit message authentication code (MAC) of an input message using an integrity key IK. KASUMI is used in a form of cipher block chaining (CBC) MAC mode.

3.2.2.1 The Confidentiality Algorithm $f8$

The initialization phase must be performed before data can be encrypted or decrypted.

The 64-bit register $A = COUNT \parallel BEARER \parallel DIRECTION \parallel 0...0$ is set according to initialization vectors $COUNT$, a time-dependant variable; $BEARER$, the bearer identity; and $DIRECTION$, indicating uplink or downlink transmission.

The counter *BLKCNT* and the first keystream block KSB0 are set to zero, while the key modifier *KM* is set to the constant value 0x555555555555555555555555555555555.

One operation of KASUMI is then applied to the register *A,* using a modified version of the confidentiality key *CK:*

$$A = KASUMI[A]_{CK \oplus KM} \qquad (3.20)$$

Once the initialization is completed, encryption/decryption operations in 3GPP are identical and are performed by the exclusive-OR of the input data with the generated keystream (KS). *LENGTH* is the number of bits in the plaintext/ciphertext, while the keystream generator produces keystream bits in multiples of 64 bits, the least significant bits of which are discarded when unnecessary.

For each integer i with $0 \leq i \leq LENGTH - 1$ we define:

$$OBS[i] = IBS[s] \oplus KS[i] \qquad (3.21)$$

Having set *BLOCKS* be equal to (*LENGTH*/64) rounded up to the nearest integer, keystream blocks (KSBs) for each integer n with $1 \leq n \leq BLOCKS$ are obtained as

$$KSB_n = KASUMI[A \oplus BLKCNT \oplus KSB_{n-1}]_{CK} \qquad (3.22)$$

where $BLKCNT = n-1$.

For $n = 1$ to *BLOCKS,* and for each integer i with $0 \leq i \leq 63$, the keystream KS[i] is obtained as

$$KS[((n-1)*64) + i] = KSB_n[i] \qquad (3.23)$$

3.2.2.2 The Integrity Algorithm f9

The initialization phase must be performed before MAC can be computed. There is no limitation on the message length on which the integrity function must be calculated.

Working variables are set as $A = 0$, $B = 0$, and the key modifier *KM* is set to the constant value 0xAAAAAAAAAAAAAAAAAAAAAAAAAAAAAAAA.

A padded string (PS), an integral multiple 64 bits long, is obtained by concatenating *COUNT, FRESH, MESSAGE,* and *DIRECTION,* and a 1 0* padding, where 0* indicates between 0 and 63 0 bits. *COUNT* is a time-dependant variable, *FRESH* is a fresh random number, *DIRECTION* indicates uplink or downlink transmission, and *MESSAGE* is the data on which the integrity function must be calculated.

$$PS = COUNT\|FRESH\|MESSAGE\|DIRECTION[0]\|1\ 0* \qquad (3.24)$$

Once the initialization is completed, the padded string PS is split into 64-bit blocks PSi, where $PS = PS_0\|PS_1\|PS_2\|...\|\ PS_{BLOCKS-1}$
For each integer n with $0 \le n \le BLOCKS - 1$,

$$A = KASUMI\left[A \oplus PS_n\right]_{1K} \qquad (3.25)$$

$$B = B \oplus A \qquad (3.26)$$

One last application of KASUMI is performed using a modified form of the integrity key IK:

$$B = KASUMI[B]_{IK \oplus KM} \qquad (3.27)$$

The 32-bit $MAC\text{-}I$ comprises the left-most 32 bits of the result, whereas the 32 right-most bits are discarded:

$$MAC - I = lefthalf[B] \qquad (3.28)$$

3.3 Bluetooth

Bluetooth [8] security services include mutual authentication and encryption but no integrity protection. Multiple publications criticize Bluetooth security, but the weaknesses reported always involve the PIN code sharing between a master and a slave, the security protocol definition, or its implementation and not the algorithms on which its security is based. This section will describe the algorithms employed for Bluetooth security, whereas protocol analysis will be discussed in Chapter 11.

Bluetooth security is based on a PIN code shared between a master and slave device. The PIN may be 1 to 16 bytes long, constant or variable. By using a PIN as a shared secret, a key hierarchy is derived: first an initialization key, then an authentication key, and finally an encryption key. Key generation functions, along with the challenge-response algorithm used for authentication, are based on a 64-bit block cipher called SAFER-SK128.

Once the encryption key has been calculated, data encryption is based on a stream cipher called $E0$ defined in the Bluetooth specifications.

3.3.1 Bluetooth Key Hierarchy and Authentication Algorithms

The block cipher SAFER-128 SK, an enhanced version of the algorithm SAFER+ [9], is the core algorithm used in Bluetooth to define

- $E\,1$, the authentication algorithm;
- $E\,2$, in its versions $E\,21$, the unit key and combination key generation algorithm, and $E\,22$, the initialization generation algorithm;
- $E\,3$, the encryption key generation algorithm.

In Bluetooth, SAFER+ is used to generate functions A_r and A'_r. A_r is identical to SAFER+, while A'_r is noninvertible and hence cannot be used for encryption. The modification to A_r used to obtain A'_r simply consists of adding in round 3 the input of round 1.

SAFER+ takes a 128-bit input to produce a 128-bit output using a 128-bit key. SAFER+ is an 8-round block cipher, needing two 16-byte subkeys for each round. The computations in each round are a composition of encryption with a subkey, a substitution, encryption with the next subkey, and a Pseudo Hadamard Transform. Nonlinearity is introduced, thanks to the use of substitution boxes.

3.3.1.1 The Authentication Algorithm $E\,1$

Authentication may be performed by either device by calculating a challenge-response function. Its inputs are

- The 128-bit challenge chosen by the verifier;
- The claimant's 48-bit address;
- The 128-bit authentication key.

The 128-bit output is divided in a 32-bit response and a 96-bit parameter called the authenticated ciphering offset (ACO).

Figure 3.4 represents the data flow for $E\,1$ computation. The output of the A_r function is xored with the random input challenge and then added byte by byte (modulo 256) with the expanded input address. This first partial result is then processed through function A'_r using a shifted value of the authentication key.

3.3.1.2 The $E\,2$ Authentication Key Generation Function

$E\,2$ is equivalent to the modified SAFER+ algorithm (i.e., A_r). The $E\,2$ function can be used in two different modes according to which key must be generated. Mode $E\,21$ shall be applied when creating unit keys and combination keys to produce a 128-bit key K using a 128-bit random value and a 48-bit device address. The specifications define which device chooses the random value and whose address is employed. Mode $E\,22$ shall be applied when creating initialization keys and eventually master keys for multicast operation. $E\,22$

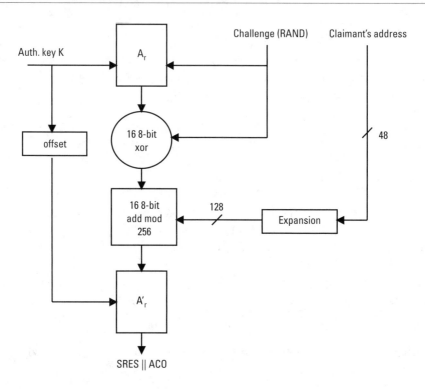

Figure 3.4 Bluetooth authentication algorithm $E1$. (After the specification of the Bluetooth System.)

produces a 128-bit key K using a 128-bit random value, the user PIN, and the device address.

3.3.1.3 The $E3$ Encryption Key Generation Function

The function $E3$ generates a 128-bit encryption key K_c by applying the A'_r algorithm to a 128-bit random value, the 128-bit link key, and the 96-bit ciphering offset COF value. Depending on whether the encryption key generated will be used for multicast or unicast traffic, the COF may, respectively, be the master device address repeated twice or the ACO result of the authentication.

Bluetooth devices may choose to use an encryption key 1 to 16 bytes long. Since the generated key is always 128 bits long, the last bits are truncated if a shorter key is needed.

3.3.2 Bluetooth Encryption Function $E0$

Bluetooth allows user information (i.e., the payload) to be encrypted, whereas the header and frame management data are sent in clear. Encryption occurs due to a stream cipher called $E0$ defined in the Bluetooth specifications.

As shown in Figure 3.5, $E0$ is the combination of three different parts. The first part allows the generation of the 128-bit payload key by combining $E0$ inputs (i.e., the encryption key K_c, the master clock, the master device address, and a random number chosen by the master). The second part is the key stream generation. The last part is the actual encryption/decryption part, in which the plaintext/ciphertext and the key stream are xored.

The keystream generator uses linear feedback shift registers (LFSRs) whose output is combined by a finite state machine as described in Figure 3.6.

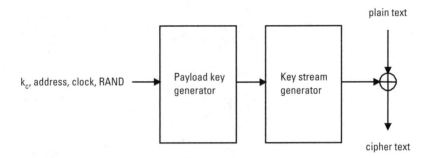

Figure 3.5 Bluetooth encryption algorithm $E0$. (After the specification of the Bluetooth System.)

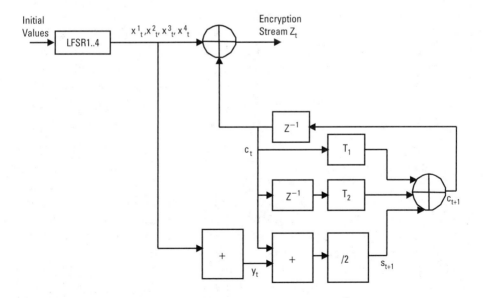

Figure 3.6 Bluetooth encryption keystream generator. (After the specification of the Bluetooth System.)

The encryption keystream is the xor of the bits in the LFSRs:

$$z_t = x_t^1 \oplus x_t^2 \oplus x_t^3 \oplus x_t^{41} \oplus c_t^0 \oplus \in \{0,1\} \tag{3.29}$$

where c^0 is an initialization value and where the whole LFSR content must be initialized before the first data encryption/decryption.

$$s_{t+1} = \left\lfloor \frac{y_t + c_t}{2} \right\rfloor \tag{3.30}$$

$$c_{t+1} = s_{t+1} \oplus T_1[c_t] \oplus T_2[c_{t-1}] \tag{3.31}$$

where

$$T_1:(x_1, x0) \rightarrow (x_1, x_0) \tag{3.32}$$

$$T_2:(x_1, x0) \rightarrow (x_1, x_0 \oplus x_0) \tag{3.33}$$

LFSR initialization must occur before encryption/decryption can start. At first the encryption key Kc is extended to 128 bits, no matter how long it initially was. Initially LFSR shift register elements and blend registers c0, c-1 are all zero. Input bits are arranged in the payload key generator according to an order specified in [8].

Symbols are generated, one per clock cycle, by shifting the bits. It is requested to generate at least 200 symbols to mix initial data before payload encryption/decryption can occur to avoid cryptanalytic attacks. It should be noted that 200 symbols have been generated after 239 clock cycles because the longest LFSR is 39 bits long; thus, 39 clock cycles must go by before significant bits fill it.

3.4   The 802.11 Standard

When the 802.11 standard [10] was originally defined, the only security services provided were authentication and encryption. The standard left to developers the burden of distributing a shared key to perform the authentication challenge response and the encryption algorithm used was Wired Equivalent Protocol (WEP), which was broken in 2001.

The medium access control security enhancements defined in the 802.11i specification [11] and ratified in 2004 define new security protocols that can be applied for authentication, encryption, and integrity protection. For compatibility with legacy devices, the old algorithms may still be employed.

3.4.1 Authentication Algorithms

The original 802.11 specification supported two algorithms for authentication: open system authentication and shared key authentication. Open system authentication is a null authentication, successful whenever the recipient agrees to use this authentication type. In shared key authentication, client authentication occurs by challenge response using WEP.

Since neither of these schemes provided adequate security, the 802.11i working group (WG) decided to support a new authentication protocol. A new protocol was not defined, it was rather chosen to support an already existent authentication protocol based on the the 802.1X specification and consequently on EAP methods. For details on 802.1X and EAP methods, the interested reader should refer to Chapter 10.

3.4.2 Encryption and Integrity Algorithms

The original 802.11 specification supported no encryption, encryption using the WEP algorithm, and no integrity algorithms. The 802.11i specification supports two new encryption algorithms combined with two integrity algorithms. Temporal key Internet protocol (TKIP) is an encryption algorithm that should be used with a message integrity check (MIC) Michael to provide integrity protection. TKIP was intended as a temporary solution for the use of legacy hardware, as its core is based on RC4, the same core on which WEP is based. For adequate security, the counter (CTR) cipher block chaining message authentication code (CBC-CCM), known as the CCM protocol (CCMP) should be used. CCMP is based on AES in CCM operation mode, therefore combining encryption and integrity.

3.4.2.1 WEP

WEP is a symmetric algorithm in which encryption and decryption occur by exoring the plaintext/ciphertext with a keystream, as can be seen in Figure 3.7. The WEP pseudorandom number generator (PRNG) is seeded with a concatenation of the encryption secret key k and an initialization vector (IV) to create a key sequence

$$\text{key sequence} = \text{WEP\_PRNG(k,IV)} \tag{3.34}$$

An integrity check value (ICV) is calculated on the plaintext and transmitted as part of the message. However, 802.11 ICV cannot be considered effective for integrity protection: it was designed to protect against data transmission attacks, and, as its computation was based on the linear function CRC-32, it is not resistant against active attacks.

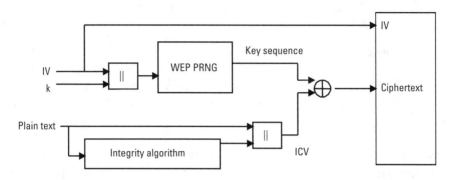

Figure 3.7 WEP encryption block diagram. (Source: IEEE 802.11i. Reprinted with permission from IEEE.)

The ciphertext and the IV in clear must be transmitted to the recipient. The 24-bit IV must be communicated for the destination device to recover the seed used by the WEP PRNG to create the key sequence. The secret key k should have previously been distributed by an external key management system. When the 802.11 specifications were drafted, a 40-bit shared key k was chosen due to U.S. government restrictions on cryptography. In an effort to improve security, WEP-2 was later defined based on the same encryption algorithm but with a 104-bit key.

WEP uses RC4 as PRNG. The stream cipher RC4 was an RSA proprietary design and was kept a secret trade until it was leaked in 1994.

Many papers were published in 2001 attacking WEP security [12–15]. Security flaws do not concern WEP's core RC4 but rather WEP's key length, the lack of a standardized mechanism for IV update, and its key mixing function in general. Attacks on WEP are reported in Chapter 12.

3.4.2.2 Temporal Key Internet Protocol and Michael

The temporal key Internet protocol (TKIP) is intended as a temporary solution providing legacy hardware with increased security compared to WEP. Just as with WEP, TKIP's core is based on RC4. It is possible to offer better confidentiality still using RC4 because WEP's problems are not due to the stream cipher itself. TKIP's security is known not to be flawless, but it is the best compromise for increased security applying a software patch on existing hardware.

Due to the lack of integrity protection in the original 802.11 specification, the MIC Michael was defined to provide integrity protection when using TKIP for encryption. Because of the design constraints of the TKIP MIC, it is still possible for an adversary to compromise message integrity, so TKIP implements countermeasures to bound the probability of a successful forgery and the amount of information an attacker can learn about a key.

Figure 3.8 shows TKIP and MIC functions. TKIP uses a 128-bit key called the temporal key. Its generation and distribution is described in Chapter 12. Besides the key, the inputs to the temporal key mixing include the transmission address TA and a TKIP sequence counter (TSC) to avoid replay attacks. Encryption is calculated on MSDUs, with the fragment block allowing the user to fragment MSDUs into multiple MPDUs if necessary. The inputs to the integrity algorithm Michael include the MSDU data, source and destination addresses, and the integrity key. For the moment the priority byte is reserved for future use.

TKIP enhances the WEP encapsulation with several additional functions:

- A new temporal mixing function, instead of the concatenation used in WEP, was defined to counteract WEP attacks.

- Encryption is performed on the data that has to be transmitted, as well as the computed MIC value. The integrity check is performed on the data, as well as the transmitter and receiver address.

- TSC update is specified in the standard. Its use avoids reply attacks.

The MIC avoids performing many attacks that were possible on WEP. As Micheal is a nonlinear function, it is impossible to succeed in bit-flipping

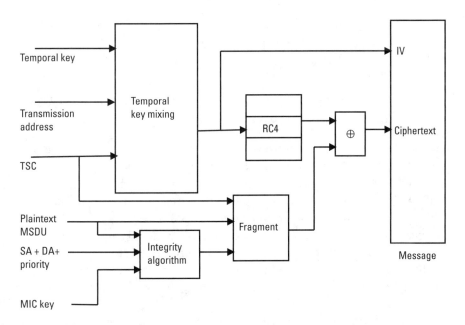

Figure 3.8 TKIP and MIC functions. (After the IEEE 802.11i specification.)

attacks, in which the attacker modifies the ciphertext and is able to build the respective correct integrity check. Data cannot be truncated or concatenated, thanks to the correct use of counters. Redirection and impersonation attacks are also prevented because source and destination addresses are included in MIC calculation.

MIC alone cannot provide complete forgery protection, as it cannot defend against replay attacks. TKIP provides replay detection by TSC sequencing and ICV validation.

Michael generates a 64-bit MIC using a 64-bit key divided in two 32-bit words $K0$ and $K1$. Michael operates on each MSDU including the priority filed, source address, destination address, and a pad.

The MIC value is computed iteratively starting with the key value (K_0 and K_1) and applying a block function b for every message word. The algorithm loop runs N times, where N is the number of 32-bit words in the padded MSDU.

The Michael block function b is a Feistel-type construction comprising additions, xors, rotations, and swap operations.

Michael's design trades off security in favor of implementability on legacy devices. If a probable active attack is detected via the failure of the Michael MIC in a received MSDU, countermeasures should be taken. The specified countermeasures include

- Logging Michael MIC failure events;

- Disabling reception for a period of 60 seconds is the rate of Michael MIC failures increases above two per minute;

- Changing the temporal key.

3.4.2.3 CCMP

The CCMP provides confidentiality, integrity, data origin authentication, and replay protection. Use of this protocol is necessary for adequate security.

CCMP is based on the CCM mode of operation of the AES [6] encryption algorithm combining CTR mode for confidentiality and cipher block chaining message authentication code (CBC-MAC) for integrity and data origin authentication. CCMP uses AES with a 128-bit key and a 128-bit block size.

CCM requires a fresh temporal key for every session and a unique nonce, which is the 48-bit packet number (PN), for each frame protected by a given temporal key. Reuse of a PN with the same temporal key voids all security guarantees.

Figure 3.9 represents the CCM protocol. The inputs to CCMP are the MPDU data, the key and key ID, as well as the PN.

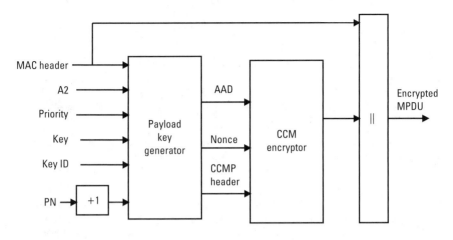

Figure 3.9 The CCM protocol. (After the IEEE 802.11i specification.)

The additional authentication data (AAD), nonce, and CCMP header construction are specified in the standard and will not be reported herein. Note that A2 is the MPDU sender address.

References

[1] Third Generation Partnership Project, "3GPP TS 03.20: Digital Cellular Telecommunications System (Phase 2+) Security Related Network Functions," http://www.3gpp.org.

[2] Third Generation Partnership Project, "3GPP TS 55.216: 3G Security: Specification of the A5/3 Encryption Algorithms for GSM and ECSD, and the GEA3 Encryption Algorithm for GPRS, Document 1: A5/3 and GEA3 Specifications (Release 6)," http://www.3gpp.org.

[3] Third Generation Partnership Project, "3GPP TS 35.202: 3G Security Specification of the 3GPP Confidentiality and Integrity Algorithms, Document 2: KASUMI Specification (Release 6)," http://www.3gpp.org.

[4] Menezes, A., P. van Oorschot, and S. Vanstone, *Handbook of Applied Cryptography*, Boca Raton, FL: CRC Press, 1997.

[5] Third Generation Partnership Project, "3GPP TS35.206: Technical Specification Group Services and System Aspects; 3GSecurity; Specification of the MILENAGE Algorithm Set: An Example Algorithm Set for the 3GPP Authentication and Key Generation Functions f1, f1*, f2, f3, f4, f5 and f5*; Document 2: Algorithm Specification." http://www.3GPP.org.

[6] FIPS PUB 197, Advanced Encryption Standard, November 2001, http://csrc.nist.gov/encryption/aes/round2/AESAlgs/Rijndael/Rijndael.pdf.

[7] Third Generation Partnership Project, "3GPPTS35.201: Technical Specification Group Services and System Aspects; 3G Security; Specification of the 3GPP Confidentiality and

Integrity Algorithms; Document 1: f8 and f9 Specification (Release 6)," http://www. 3GPP.org.

[8]   Specification of the Bluetooth System, Core Package version 2.0 + EDR, 4 November 2004.

[9]   Massey, J. L., G. H. Khachatrian, and M. K. Kuregian, "Nomination of SAFER+ As Candidate Algorithm for the Advanced Encryption Standard (AES)," First Advanced Encryption Standard (AES) candidate conference, Ventura, CA, Aug. 20–22, 1998.

[10] Information Technology—Telecommunications and Information Exchange Between Systems—Local and Metropolitan Area Networks—Specific Requirements, Part 11: Wireless LAN Medium Access Control (MAC) and Physical Layer (PHY) Specifications, ANSI/IEEE Std 802.11, 1999 Edition.

[11] IEEE Standard for Information—TechnologyTelecommunications and Information Exchange Between Systems—Local and Metropolitan Area Networks—Specific Requirements Part 11: Wireless Medium Access Control (MAC) and Physical Layer (PHY) Specifications: Amendment 6: Medium Access Control (MAC) Security Enhancement, ANSI/IEEE Std 802.11, 2004 Edition.

[12] Borisov, N., I. Goldberg, and D. Wagner, "Intercepting Mobile Communications: The Insecurity of 802.11," draft, http://www.isaac.cs.berkeley.edu/isaac/wep-draft.pdf.

[13] Fluner, S., I. Mantin, and A. Shamir, "Weaknesses in the Key Scheduling Algorithm of RC4," *Selected Areas in Cryptography*, 2001.

[14] Stubblefield, A., J. Ioannidis, and A. Rubin, "Using the Fluner Mantin and Shamir Attack to Break WEP," AT&T Labs Technical Report TD-4ZCPZZ, August 6, 2001.

[15] Rivest, R., "RSA Security Response to Weaknesses in the Key Scheduling Algorithm of RC4," http://www.rsasecurity.com/rsalabs/technotes/wep-fix.html.

4

Hardware Security

4.1 Introduction

Today, there are new security challenges with the deployment of emerging services on mobile platforms, such as banking, digital media services, wireless commerce, networked gaming, and third-party software downloads.

With the convergence of these services, the vulnerabilities of existing mobile and wireless platforms constitute a major risk to existing network revenue. The deployment of these new services must be done without disruptions to the operator's network infrastructure or jeopardizing consumer's private information and premium content stored on the mobile platforms.

Making a mobile platform safe from malicious attacks has consequences for hardware and software design, as well as the physical attributes of the design. The best-protected mobile systems must have security measures designed in from the outset, starting with the specification for the processor or CPU core. A comprehensive security solution that combines software and hardware can help support platform trust operations, security protocols, access control mechanisms, and protection of private data and valuable content. This security solution should provide the infrastructure where only well-behaved applications and services can thrive, while rogue applications and viruses would be quarantined. Such a solution could mitigate the risk to wireless networks and ensure existing revenue for wireless service providers.

The level of security provided by security hardware on a mobile platform can be extremely high. However, unless specific manufacturing steps are taken to guard against physical attack, no secure system can be guaranteed to be unbreakable against very sophisticated and sustained attacks. The mobile platform designer's goal should be to raise security to the right level when considering future threats while not forgetting the economic and practical aspects of systems implementation.

This chapter describes the hardware aspects of mobile security and delves into existing and emerging hardware protection technologies.

4.2 Threats Addressed by Hardware Protection

The list of potential threats on mobile platforms is quite long. That list includes corruption of the platform's internal resources, unauthorized access to private data or services provided by the platform, cloning, and theft of valuable content.

Also, mobile operators and manufacturers require some level of protection of the mobile platform from illicit software and hardware modifications. Software modifications may be spread over the network through viruses. If an attacker has physical access to the mobile platform, it could also lead to the memory being re-programmed (memory-reprogramming attack). A trusted-boot ROM capability can protect the platform from this family of attacks using strong cryptographic checks to validate the integrity of the platform software.

Furthermore, strong protection is needed to guarantee that the user's personal data, credit card information, and stored value are reasonably protected from attack. Hardware-based secure storage can help protect user data against observation or modification (observation and modification attacks) and network-operator data using strong encryption with integrity checks. This can be designed to allow large amounts of data to be stored in system memory without risk of observation or risk of modification without detection.

Network operators have a very strong desire to protect the IMEI from being modified, for it is the identity of the mobile phone and, if modified, the stolen mobile platform can be given an IMEI replacement (identity-spoofing attack). Therefore, it is critical that the IMEI is well protected while it is stored as well as when in use. Hardware protection can help enable the protection of the IMEI at all times, even when the IMEI is being used or is being transferred between subsystems, by using encryption and/or physical partitioning.

Protecting multimedia content, such as video and audio files, is often required by service providers. Security hardware can provide strong mechanisms to protect the high-value content on the mobile platform and to ensure that the DRM is not violated. Content that is temporarily stored on the mobile platform can be protected by strong encryption, and the access control policies can prevent unauthorized access to the keys needed to decrypt the content. The specific management information governing the use of a particular file is protected by encryption and integrity checks. Security hardware can allow the software that enforces the DRM policy to be completely checked for integrity at boot time and to be retested each time the DRM application is launched.

An unauthorized program should not issue privileged system instructions (such as instructions to change the status of the system or to initiate I/O), nor should it become authorized except through controlled system hardware or software interfaces. Thus, it should not interfere with the OS or applications. Hardware-based execution isolation can provide sufficient isolation of such an unauthorized program.

Also, isolation can be provided by storage protection keys and execution states. The hardware can further isolate the user programs into different address spaces. Each address space has the ability to read common system storage, but it can neither read nor write the nonshared storage belonging to another address space unless allowed to do so by an authorized program. Hence, unauthorized programs cannot interfere with programs in another address space. Programs in different address spaces must share data, and the OS should provide mechanisms to allow such sharing in a safe and controlled way.

Hardware-based integrated cryptography can enable a mobile platform to encrypt and decrypt data, generate and manage cryptographic keys, and perform other cryptographic functions dealing with data integrity and digital signatures. Such coprocessors can have a tamper-responding design, a feature that is hard to support in software-based cryptographic APIs.

On mobile platforms, private keys used by applications to authenticate themselves and to digitally sign data and communications must be protected from loss (key-spoofing attacks). The compromise of those keys could lead to loss of trust as well as financial losses. Hardware-based cryptographic engines can be highly secure, with the master encryption keys stored within the hardware boundary and used, in turn, to encrypt working keys.

Side-channel attacks include timing analysis attacks, power analysis attacks, and electromagnetic analysis attacks. Timing analysis attacks allow inferring information on the data or secret values due to a dependency between code execution time and data being processed. A timing analysis attack may simply watch for the length of time a cryptographic algorithm requires. However, most timing analysis attacks watch data movement into and out of the CPU or memory on the hardware running the cryptosystem or algorithm. By observing how long it takes to transfer key information, it is possible to determine how long the key is. In some cryptographic implementations, internal operational stages may provide partial information about the plaintext or key values, and some of this information can be inferred from observed timings. Physical security of hardware can be used to reduce the risk of malicious installation of devices that can provide time analysis, such as microphones and micromonitoring devices.

A power analysis attack is a form of side channel attack in which the attacker studies the power consumption of a cryptographic hardware device (such as a smart card). This attack can yield information about what the device is doing and even some key material. In SPA the power consumption of a single algorithm execution is analyzed, whereas in DPA information is deduced by performing statistical analysis of power consumption curves over several executions of the same algorithm. Using a hard-wired hardware cryptographic device, where power consumption can vary very little due to its construction, can prevent power analysis attacks.

A DMA attack is a hardware attack that involves repurposing built-in bus-mastering hardware to be not successfully executed. This form of attack cannot be mitigated without the aid of a complete solution comprising of hardware, processor, software, and hypervisor.

On Windows-based mobile platforms, many trojans and viruses rely on several types of Basic Input Output System (BIOS) modification attacks. Hardware-based secure-boot solutions can prevent attackers from successfully executing a BIOS-modification attack (see Section 5.3.2). In addition, secure-boot can ensure that a trojan cannot be inserted during the boot before an administrative user has been successfully authenticated.

4.3 Hardware Security Solutions

4.3.1 Smart Card Technology

4.3.1.1 Introduction

A smart card [1] is a plastic card with a chip inserted in it. Its invention dates back to the 1970s. The plastic support most often contains printed information on the cardholder and advertising from the issuer. Optionally, the plastic support can also carry magnetic stripes and bar codes as needed, depending on the card's specific use.

A smart card is itself a mobile device but not a stand-alone device, as it must be connected to a terminal through a reader to be powered on and to employ the reader's interfaces for communication. Smart cards can be contact cards or contactless card. In the former, the chip is inserted under the micromodule, which is a visible metal rectangle with eight contact pins (see Figure 4.1).

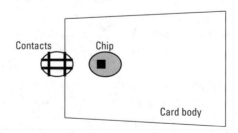

Figure 4.1 Contact smart card.

Contact cards need to be inserted inside a reader, either ISO or USB. It is the reader that supplies power to the card through one of the pin contacts. In contactless cards, the chip is embedded in the plastic and communicates with the reader by radio frequency (see Figure 4.2). Contactless cards only need to be swiped over the contactless reader to communicate with it; power is provided by the reader via electromagnetic induction.

Portability and tamper resistance are the main smart card features, where tamper resistance is defined as the card's capacity to resist to invasive attacks, fault attacks, and side channel analysis. Smart cards are mainly used to memorize and process sensitive user data. Compared to most hardware available on a large-scale basis, smart card manufacturing involves an extra step: personalization. Personalization is a highly optimized process in smart card manufacturing, and it allows the loading of unique values per card (e.g., card identifiers or cryptographic keys).

Smart cards are divided into memory cards and processor cards. The former are low-cost devices deployed to store data (e.g., remaining credit in telephone cards, vending machines, or metro tickets). The latter are able to store and process data and can be used for a variety of banking, telecommunications, identity, and other applications.

Microprocessor smart cards today can have an 8-bit or a 32-bit CPU. Memory capacity is about 2 KB of RAM, 48 to 64 KB of ROM, and 8 to 32 KB of EEPROM or FLASH. To function, smart cards need an external 3- to 5-volt power supply and an external 1- to 5-MHz clock, but to overcome external clock limitations, most chips run an internal clock up to 30 MHz. I/O allows a serial half-duplex rate of 9.6 to 30 Kbps, ISO 7816/3 rules reader to smart card communication.

New generation high-end smart card hardware, with 32-bit CPUs, is expected to be rolled out in 2007. Memory capacity is expected to be 16 KB of

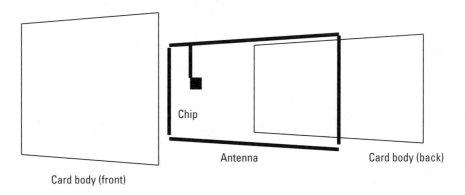

Card body (front)

Chip

Antenna

Card body (back)

Figure 4.2 Contactless smart card.

RAM, 256 KB of ROM, and more than 128 KB of EEPROM or several megabytes of FLASH. The internal clock should reach 50 MHz, and the I/O should accommodate a full-duplex rate of 1.5 to 12 Mbps. Also, next generation cards will not require a smart card reader but will support direct USB, MMC, and eventually contactless connectivity.

Since protection of sensitive data and sensitive execution is the smart card's reason to be, all propose cryptographic algorithms. The algorithms available depend on application needs and card capacity. Most cards offer standard secret key algorithms. High-end cards may offer public key algorithms. To enhance performance, dedicated cryptographic coprocessors, such as for AES, Rivest-Shamir-Adlemann (RSA), or elliptic curve cryptography (ECC), may be available. At the client's request, proprietary algorithms may also be developed.

4.3.1.2 New Generation Smart Cards

In this section, we will concentrate on processor smart cards, as memory smart cards are commodity devices using a basic technology that evolves very little besides increases in capacity. Processor smart cards, which we will now refer to simply as smart cards, exist with different operating systems (OSs), cryptographic features, and security features.

The OS is one of the main smart card elements. Initially dedicated to a single chip, it is today chip independent. It relies on hardware features typically found in personal computers, including communication drivers, a universal asynchronous receiver transmitter (UART), memory banking for enhanced addressing capacity, a memory management unit (MMU), timers, and interruption handlers.

Time-to-market constraints require software to be portable and flexible, so the programming language used evolved from assembly to Java. Java card is a dedicated instruction set for smart cards (see Figure 4.3). Just as Java does for computers, Java card allows any Java application to run independently of the underlying smart card chip or OS. Moreover, to distinguish them from native

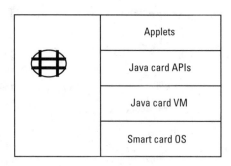

Figure 4.3 Java smart card.

cards, Java cards are called open cards because of their capacity to support the download of new services and commands during the card life.

Another asset of Java is the intrinsic security it brings to smart cards, thanks to the secure execution environment it sets up, with a firewall between different applications in the same card so they can function separately and independently.

Java Card supports the main security features inherited from the Java language, plus

- Transactions are atomic.
- A firewall separates applets in the same card.
- Security and cryptography classes support encryption and decryption, signature generation and verification, message integrity, random number generation, and PIN and key management.

Java Card uses a *split virtual machine*, meaning that the executable code is prepared off-card and then executed on the on-board virtual machine to enhance smart card performance and comply with the limited resources available. The off-card component is a converter and verifier. It accepts Java class files, performs linking and resolves references, and then produces a CAP file that is loaded onto a Java card product. The on-card part of the Java virtual machine executes the CAP file code and enforces security.

Java Card 2.2.1 [2–4] addresses all major smart card markets, including mobile telecommunications, identity, finance, pay TV, and transportation. Java Card 2.2.2, the current version, contains specific enhancements for contactless cards and e-passport applications. The next generation Java platform will be backward compatible. It should support enhanced Java language features an enhanced cryptographic toolkit. The connection-based model will be servletlike, and the platform will be able to initiate connection-based requests to other servers.

New generation smart card prototypes have been publicly presented to the industry [5]. In 2008, it is expected that smart cards will enable end users to access and manage private data directly from their Web browsers. Reciprocally, client applications will be able to access Web services offered by the smart card. Thanks to the Web interface, card issuers will also be able to remotely manage the smart cards on the field. New generation prototypes run enhanced virtual machines, have multithreading capabilities, and embed automatic garbage collection.

4.3.1.3 Smart Card Security

Absolute security does not exist, that is why smart cards are defined as tamper resistant not tamper proof. The appropriate level of security must be established, depending on the value of the data that should be protected. A compromise between the security cost and the remaining risk of the smart card being broken must be balanced. Risk management allows estimating at best the security that should be implemented to protect the system against identified attack scenarios.

Smart cards can store sensitive data, protecting it from outside world access, and can process it without leaking any information. Smart cards are always part of a system, so to grant overall system security, architects cannot focus on the smart card alone. In a chain, the highest security level is equivalent to that of the weakest link.

Smart card security mechanisms are designed to offer confidentiality and integrity against invasive attacks, side channel analysis, and fault attacks. Software attacks such as design and implementation errors or flaws in cryptographic protocols are also possible, as with any electronic device.

Invasive attacks, also known as physical attacks, irreversibly modify physical properties of the chip while aiming at capturing information stored in memory areas or flowing over the data bus. Delicate chemistry and electronic manipulation is needed to disconnect circuits, override sensors, or dissolve shields. These attacks are very costly, both in required equipment and technical knowledge. Similar techniques can be applied to any electronic device, but smart cards have more hardware countermeasures than most general-purpose hardware on the shelf. The first countermeasure consists of embedding the complete system, including the CPU, memories, and peripherals, in a single chip. Moreover, the design usually includes additional security features, such as protection shields, glue logic design, bus and memory encryption, and scrambling. This makes it very difficult to locate functional blocks and to retrieve information by analyzing the chip structure. Smart card chips are made of multiple layers so that sensitive components can be hidden in buried layers. Voltage, light, temperature, and clock frequency sensors can also be activated to prevent the chip from operating in abnormal conditions.

Side-channel analysis consists of monitoring a device signal such as the processing duration, power consumption, electromagnetic radiation, and radio-frequency emission to infer information about a secret data processed during the acquisition's period of time. This noninvasive technique requires hardware devices to monitor the targeted signal and knowledge of electronics, cryptography, signal processing, and statistics. Side-channel analysis includes:

- Timing analysis;
- Power analysis;
- Electromagnetic analysis.

Timing analysis allows the inferring of information on the data or secret values due to a dependency between code execution time and data being processed. An effective countermeasure consists of developing code in which timing does not reflect processed data.

Power analysis encompasses simple power analysis (SPA) and differential power analysis (DPA) and consists of assuming information on the data or secret values due to a dependency between the chip power consumption and data being processed. In SPA the power consumption of a single algorithm execution is analyzed, whereas in DPA information is deduced by performing statistical analysis of power consumption curves on several executions of the same algorithm. Countermeasures involve an implementation in which power consumption does not reflect processed data. Noise and clock delays can also be inserted to complicate curve interpretation and synchronization.

Electromagnetic analysis is based on the same techniques as power analysis; the only difference is that information is hidden in the radio frequency signal. Compared to power analysis, electromagnetic analysis allows the targeting of specific chip areas, as bus lines or memory areas, which leak information. On the other hand, measuring the chip power consumption is easier than measuring electromagnetic radiation over a specific spot. Countermeasures against electromagnetic analysis mainly involve use of protective shields.

Fault attacks are the latest class of attacks that emerged. They rely on a physical perturbation of the standard environmental conditions. The fault induced by the physical perturbation will cause an abnormal behavior of the chip. In certain cases, this will allow disclosing secret data or enable actions that are normally denied. For example, the comparison between the chip behavior with or without the fault may permit the recovery of information on values being processed. Likewise, a fault may consent execution of forbidden operations, such as accessing protected memory areas or installing code without performing security verifications. The fault may be induced in different ways: through an electromagnetic field, a power glitch, or a laser beam. It is very difficult to master the effect of the fault as well as the moment it will appear, but research in this field is improving the technique. The mostly adopted countermeasure consists of the use of sensors to detect abnormal operating conditions. Results can also be computed twice and compared to detect an eventual fault.

4.3.1.4 Smart Card Applications

Smart cards can be applied in any business field, including financial services, mobile communications, identity, health, and transport services.

Most credit (and debit) cards are magnetic strip cards, but these can be easily forged so the fraud rate is relatively high. To improve security, credit card companies, namely Europay, Mastercard, and Visa (EMV), developed a specification for integrated circuit card payment systems. Surveys prove that fraud rate is much lower on smart banking cards than on magnetic strip cards. In the past, attacks were successfully performed on smart card credit cards because of a flaw in the security protocol used [6]. EMV specifications offer different authentication solutions. They became a defacto standard, and more countries are migrating toward this standard.

In magnetic strip cards, user identification and bank account values are not strongly protected. User identity is verified by requesting a manual signature, but this is not always checked and can be easily forged. User ID is sometimes requested, but it is not mandatory. Finally, before registering the transaction, bank clearance has to be requested.

In transactions with smart card credit cards, the first step will consist of asking the user for her PIN to prove she is the card owner. User identity and bank account information is securely stored inside the chip. Low-risk transactions, can be performed offline. Authentication will be performed between the merchant's terminal and the smart card, and then the transaction will be registered. Certain transactions may still require online bank clearance, in which case the protocol deployed is the same one as that for magnetic strip cards.

Telecommunications is an area where smart cards are largely widespread. Besides their use for storage of credit to use in fixed phone booths, which can be considered a use case example for electronic payment, their main application is as the link between the mobile phone operator and the subscriber in GSM and 3GPP. Details are provided in Chapter 14.

Electronic identity is a fast-expanding sector, due to the need for improved safety after recent events. Governments want to upgrade identity verification means from paper, which is in use today, to systems including bar codes, holograms, biometric recognition systems, and electronic chips. The International Civil Aviation Organization (ICAO) is coordinating international efforts to define compatible identification means. It has defined a report to provide guidance and advice to states regarding the application and use of contactless integrated circuits in machine-readable travel documents [7].

4.3.2   Trusted Platform Module Technology

4.3.2.1   Introduction

The Trusted Computing Group (TCG) has produced open specifications for a security chip, called Trusted Platform Module (TPM) [8, 9] and the related software interfaces (see Figure 4.4). The TPM specifications define minimal hardware-based security requirements for client-side security. The TCG specifications provide two main security functions: secure storage of signature and encryption keys and system software integrity measurement. The TPM's secure

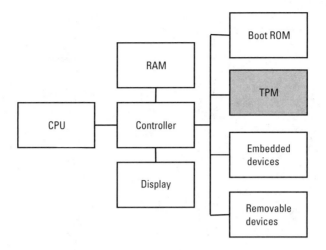

Figure 4.4 TPM configuration.

storage can be used to protect an individual's RSA authentication private key from theft or disclosure. The TPM's integrity measurement can be used to detect software compromise, and to lock down the use of protected keys and data if a compromise is detected.

The TCG specifications [8, 9] define a *trusted platform* as a platform consisting of trusted hardware and software, and the integration of external certification authorities for enabling cryptographic-proof mechanisms. The trusted hardware is the TPM, which is the central hardware security device where all elementary operations are securely managed. The TPM is typically implemented as a discrete chip, usually installed on the motherboard of a device, that communicates with the rest of the system using a hardware bus. The TPM supports the following cryptographic functions: hashing, random number generation, asymmetric key generation, and asymmetric encryption and decryption. Each single TPM on each platform has a unique signature initialized during the silicon manufacturing process and must have an owner before it can be utilized as a security device.

4.3.2.2 Architecture

Components The main security functions (see Figure 4.5) supported by the TPM are:

- *Protection of key material:* Various key categories can be stored in a protected way in the TPM. The access method is selected according to the key type.

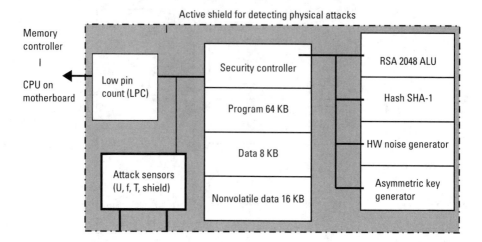

Figure 4.5 TPM architecture.

- *System authentication:* Authentication and validation of the platform to third parties.

- *Communication of the system's security status (attestation):* Trusted communication of the security-relevant configuration. This configuration is defined by the platform user.

- *Random number generator:* Generation of hardware-based random numbers for secure key generation.

- *File sealing:* Binding of data to the system configuration and signing of the data when storing with the hash value of the configuration. Access to the data is then only possible if the configuration remains unchanged.

- *Secure saving of configuration changes in the platform configuration register:* Status changes are detected, safeguarded by the SHA-1 hash algorithm.

The main cryptographic and security hardware components (see Figure 4.5) of the TPM are

- Specialized cryptographic arithmetic unit for fast computation of RSA cryptography up to 2,048 bits;

- Key generation for RSA keys up to 2,048 bits;

- Hardware hash unit for the SHA-1 algorithm;

- Genuine hardware noise generator as input for key generation;

- Internal processor with the appropriate hardware for computing the critical functions (e.g., RSA with the secret key part) on a trusted basis in a secure environment;

- Monotonic, protected counters used to prevent replay attacks;

- Nonvolatile memory (EEPROM) to retain data even when the system power is switched off;

- Sensors and internal security structures (e.g., active screen over the top wiring layer of the chip) in order to detect physical attacks and counteract them;

- TPM self-test functions.

Firmware and Software   Besides the hardware components, the TPM has internal firmware that implements the interface protocol defined by the TCG specifications to the overlying layers of the TCG software stack and uses the hardware functions for this purpose. In addition, this firmware also checks and administers the various security sensors and reacts appropriately to detected physical tampering or alterations to the chip or its environment. The correctness of the implementation is checked and confirmed by an independent test enforced by a complex certification process.

The TPM's functions are provided to the OS using the TCG software stack (TSS) (see Figure 4.6). The TSS consists, at the lowest level, of the hardware- based device driver, which initializes the interfaces and exchanges data with the TPM. The next higher level consists of the system service, which consists of the TPM device driver library. This coordinates and manages multiple accesses to the TPM; the TSS core services, which converts the abstract API commands to the data stream for the TPM; and the TSS service provider, which reads the system service for remote access.

The TCG specifications define three core TPM objects: the core root of trust for measurement (CRTM), the trusted platform support service (TSS), and the initial program loader (IPL). The CRTM consists of the routines executed at the start of booting of the platform, before the OS is available, in order to achieve secure startup conditions. This is accomplished by measuring and monitoring the integrity of the boot operation, whereby hash values of the critical parts are formed and then provided to the TPM for checking. The TSS provides the OS with a standardized high-level API to the TPM via which it handles the security functions of the OS or applications. The IPL is the link between the BIOS and the OS and ensures the integrity of the OS.

The Static Root of Trust for Measurement (SRTM) mode of the TPM protects against software-based attacks which are the majority of the known

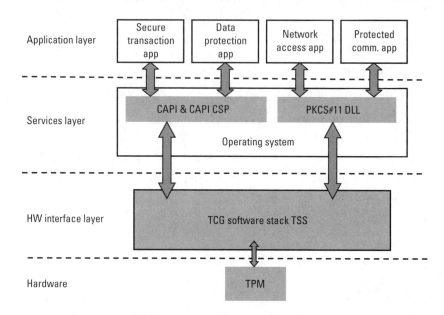

Figure 4.6 TPM software stack.

attacks. However, the SRTM mode of the TPM is susceptible to hardware attacks. A successful hardware attack typically requires expensive hardware that can be difficult to obtain and significant expertise to carry out. This mitigates the risk to a very small segment capable of this level of attack.

Key Management TPM-enabled platforms are capable of creating cryptographic keys and encrypting them so that they can be decrypted only by the TPM. This process, which can help protect the key from disclosure, is called wrapping or binding. The master wrapping key stored within the TPM is called the storage root key (SRK). The storage of this key on the TPM ensures that the private portion of the key is never exposed. Platforms with a TPM are also capable of creating a key that has not only been wrapped, but also tied to certain platform measurements such that the key can only be unwrapped when those platform measurements have the same values that they had when the key was created. This process is called sealing the key to the TPM. Decrypting this key is called unsealing. The TPM can also seal and unseal data that is generated outside of the TPM.

The main TPM keys are as follows (see Figure 4.7):

- *Endorsement key (EK):* The manufacturer generates this 2,048-bit private/public key pair in the TPM chip at the end of its fabrication.

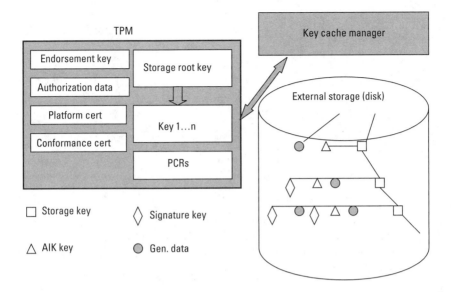

Figure 4.7 TPM key management.

The EK is stored in such a way that the private key can no longer be read out but can only be used internally in the TPM. The EK is further protected by a special certificate. The manufacturer thereby confirms electronically that this TPM has been produced in a trusted process by an inspected manufacturer and meets the requirements of the TCG specification. The trustworthiness of the entire TPM system is based for the most part on this process and the uniqueness of the EK. The user must trust the manufacturer that the private part of the key is not stored anywhere and that it is not accessible to anyone else.

- *SRK:* The SRK forms the root of a key hierarchy in which other lower order keys, but also data (blobs), are securely stored. Their trustworthiness therefore depends on the SRK. The SRK is automatically generated by the owner in a "take-ownership" operation. If the owner of a TPM gives up this ownership, this also deletes the SRK and makes all the keys protected by it completely unusable.

- *Attestation identity key (AIK):* The AIKs are derived from the SRK and can also be subsequently created or deleted based on their use. In the TCG context, attestation refers to both authentication and integrity. A platform can have multiple AIKs for each user. The AIKs can be used for server authentication, platform-bound digital contents (e.g., DRM), anonymous identities in procurement, and tender platforms.

The TCG specifications define several cryptographic certificates that are also stored in the TPM:

- *Endorsement certificate:* This certificate confirms that the TPM originates from a trusted source. It contains the public key of the EK and is used for deriving the AIK.

- *Platform certificate:* This certificate is introduced by the motherboard or platform manufacturer and confirms that a valid TPM has been mounted on a correct platform. This certificate is also used for deriving the AIK.

- *Conformance certificate:* This certificate is issued by a test laboratory and confirms that the security functions of the TPM and motherboard have been positively checked and are compliant with the protection profile of the TCG.

4.3.3 TrustZone Technology

4.3.3.1 Introduction

TrustZone [10] can be either built in to the Advanced Risk Machine (ARM) CPU's instruction sets or provided in security modules by other vendors (e.g., Trusted Logic). The architectural aspects of TrustZone are implemented starting with the ARM11 CPUs; however, the security modules bring the TrustZone framework to all the ARM CPUs through a set of common application programming interfaces (APIs).

The TrustZone solution provides several security functions, such as platform identification and authentication, identity, key and certificate management, low-level cryptography, I/O access control, safe data storage, smart card access, and code-integrity checking.

4.3.3.2 Architecture

The TrustZone solution consists of a hardware-enforced security environment that provides code isolation and software, which establish the fundamental security services and interfaces to other elements in a trusted chain such as smart cards. The code-isolation capability separates a nonsecure execution environment from a trusted and certifiable secure environment (see Figure 4.8).

TrustZone operates by enforcing a level of trust at each stage of a transaction, such as system boot. By executing secure commands within a trusted execution environment, the trusted code can, for example, protect the decryption of messages using the recipient's private key or verify the authenticity of the signature based on the sender's public key (see Figure 4.9).

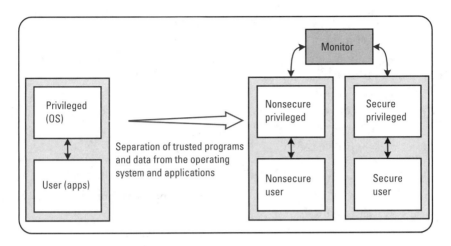

Figure 4.8 TrustZone's partitioning.

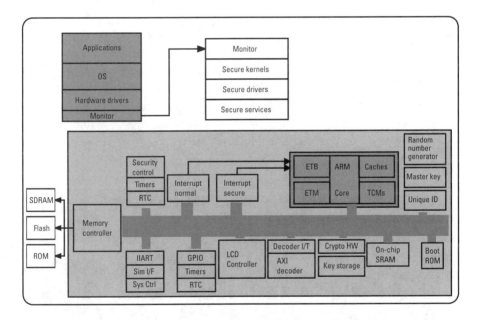

Figure 4.9 TrustZone architecture.

In order to determine whether the system is operating in the secure or nonsecure environments, TrustZone uses a secure monitor mode that controls switching between the two environments. A secure monitor interrupt instruction provides the main route to change environments.

Furthermore, TrustZone provides additional security capabilities, including secure on-chip boot ROM to configure the system, on-chip nonvolatile memory for storing device or master keys, and secure on-chip RAM used to store and run trusted code or to store secrets.

To enable security within the OS, TrustZone can provide integrity checks against attacks in three ways:

- It can verify that the OS is unaltered before booting it.
- It can verify that critical paths are unaltered during run time.
- It can safely execute a restricted set of approved functions remotely from the main OS.

As shown in Figure 4.10, software applications can be deployed three different ways in TrustZone:

- A nonsecure application runs directly on the OS in the nonsecure environment.
- A secure application A can go through the OS, which calls the access driver that switches to the secure environment. When the kernel receives the request, the API manages the secure key storage.
- For a secure application B, the secure operation can be enabled directly through the TrustZone trusted interpreter, bypassing the OS completely.

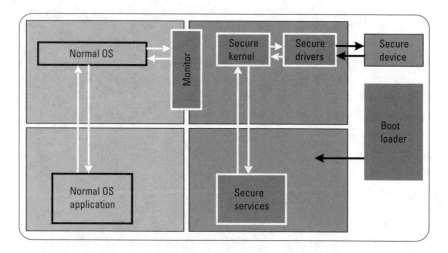

Figure 4.10 Secure software elements.

4.3.4 Wireless Trusted Platform Technology

4.3.4.1 Introduction

Several researchers studied the use of the TPM on mobile platforms [11–13] for hardening security software, protocols, and hardware. Recently, Intel introduced its wireless trusted platform (WTP) [14, 15], which is designed to provide platform trust and several additional security services incorporated in the Intel PXA27x processor family for mobile devices. The WTP solution is based upon the concepts developed by the TCG industry forum [8] and offers capabilities that include trusted-boot, protected storage of private information and keys, cryptographic acceleration, and key management. The physical protection of critical WTP components is done by placing the entire system on a chip (SoC), which can protect them from removal, replacement, and tampering.

Trusted boot checks the integrity of the platform and the applications loaded into the memory. It is capable of recognizing virus activities, malicious software, and alterations of platform configurations.

Protected storage allows the user to store sensitive information on a nonvolatile memory, which is protected by a strong cryptographic algorithm. Protected storage also provides integrity checking, which can protect against tempering with the data.

The OS and user applications can access WTP through cryptographic APIs (see Figure 4.11).

4.3.4.2 Architecture

The Intel WTP provides a set of hardware and software capabilities that provide the basis for a mobile trusted computing environment. Additional security components such as virus scan software can be built upon, and take advantage of the underlying WTP security architecture. As shown in Figure 4.12, the WTP architectural components mainly include

- Intel trusted-boot ROM;
- Intel wireless trusted module;
- Protected storage;
- Physical protection.

Intel Trusted-Boot ROM   The Intel trusted-boot ROM is the WTP component that validates the integrity of the platform and boots the platform into a known configuration. Trusted boot is an active part of the security solution during all stages of a product's life cycle, from manufacturing through the sale and use of the platform by consumers. It is invoked whenever power is applied or when commanded by the OS. The Intel trusted boot ROM is first invoked during

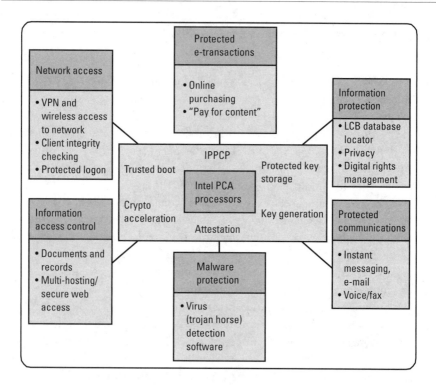

Figure 4.11 WTP usages.

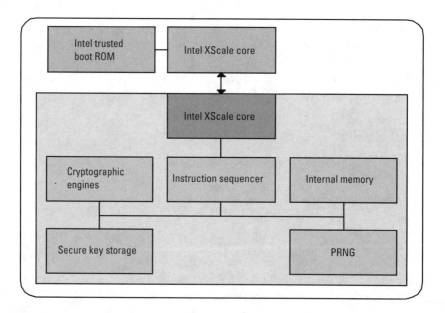

Figure 4.12 WTP architecture.

device manufacturing. As part of the manufacturing boot, a device may also be loaded with cryptographic keys used for digital signature verification of the code objects and the secure enabling of other functions that require asymmetric keys. At the manufacturing stage, trusted-boot is designed to validate the integrity of the code and keys, and it authenticates that the objects being loaded have been signed by the manufacturer. The manufacturer must format the code objects to be consistent with the formatting expected by the trusted-boot software.

Once the platform is deployed, a power-on event initiates trusted boot, which now can validate the integrity of the software code objects on the platform and detect any modification to the platform software configuration originally loaded by the manufacturer.

During trusted-boot, the trusted-boot code performs a cryptographic measurement of the platform's code objects and compares the measured value to a known good value. The measured value is also stored and can be presented to some entity at a later time in order to verify the state of the platform at boot-time. Trusted-boot is based on TCG's transitive trust model: it is initiated by the trusted-boot code, then the trusted-boot validates the integrity of other software objects, and since those objects are included inside the trust boundary, their functions and capabilities can be used to further extend the trust boundary until the entire platform has been checked and is trusted. The trusted-boot code is stored in memory that cannot be modified or bypassed.

Intel Wireless Trusted Module   This module provides a safe area to process secrets and includes a suite of cryptographic engines to support a core set of cryptographic primitives. The primitives include random number generation, symmetric and asymmetric cryptography, key creation, key exchange, digital signature operations, hashing, binding, and a monotonic counter. These security operations, which are used to construct higher level security functions, are atomic, and their intermediate results are not revealed and are not modifiable by agents outside the module. Also, these operations cannot be monitored or altered by the application processor. Attestation uses this module to provide information about the operating environment on a mobile platform. The attestation can either represent a measurement of the device at boot time or at any time after boot.

The cryptographic algorithms and functions supported by this module include AES electronic code book (ECB) mode, CBC and countermode, RSA, secure hash algorithm 1 (SHA-1) and SHA-1-based hashed message authentication code (HMAC), Random Number Generator (RNG), and digital signature creation and verification.

The module can also provide protected storage in system flash. Protected storage provides secure nonvolatile storage of secrets, such as passkeys, cryptographic keys, DRM data, and e-cash.

Physical Protection   In order to mitigate these threats of critical security components being bypassed, removed, or replaced, a physical protection mechanism is a key component of the WTP architecture. The WTP security hardware is integrated in a single device (SoC), and its discrete components are packaged into a single physical package.

References

[1]   Rankl, W., and W. Effing, *Smart Card Handbook,* 2nd ed., London: John Wiley & Sons.

[2]   Java Card 2.1 Application Programming Interface, Sun Microsystems, June 1999.

[3]   Java Card 2.1 Runtime Environment, Sun Microsystems, June 1999.

[4]   Java Card 2.1 Virtual Machine Specification, Sun Microsystems, June 1999.

[5]   Ravishankar, T., A. Sharma, and L. Lagosanto, "Java Card Technology and Tomorrow's Security Architectures," Mobility and Devices Session TS-7136, JavaOne Conference 2005, https://jsecom16.sun.com/ECom/EComActionServlet;jsessionid= 959B86B6B379E E60EC6D91E866EA2B5D.

[6]   Jessel, S., "Credit Card Whistleblower Sentenced," BBC News, February, 25, 2000.

[7]   ICAO doc 9303, Annex I, Use of Contactless Integrated Circuits in Machine Readable Travel Documents, version 4.0, May 2005, http://www.icao.int/mrtol/.

[8]   Trusted Computing Group, http://www.trustedcomputinggroup.org.

[9]   Pearson S. (ed.), *Trusted Computing Platforms: TCPA Technology in Context,* Englewood Cliffs, NJ: Prentice Hall, 2003.

[10]   Alves, T., and D. Felton, TrustZone: Integrated Hardware and Software Security— Enabling Trusted Computing in Embedded Systems, http://www.arm.com/pdfs/TZ%20 Whitepaper.pdf.

[11]   Aissi, S., et al., Trusted Mobile Platform Protocol Specification, http://xml.coverpages. org/TMP-ProtocolV10.pdf.

[12]   Aissi, S., et al., Trusted Mobile Platform Software Architecture Description, http://www. trusted-mobile.org/TMP_SWAD_rev1_00.pdf.

[13]   Aissi, S., et al., Trusted Mobile Platform Hardware Architecture Description, http://xml. coverpages.org/TMP-HWADv10.pdf.

[14]   Intel PXA27x Processor Family, http://www.intel.com/design/pca/prodbref/253820.htm.

[15]   Intel Wireless Trusted Platform: Security for Mobile Devices, http://www.intel.com/ design/pca/applicationsprocessors/whitepapers/30086801.pdf.

5

Software Security

5.1 Introduction

In this chapter we describe software security approaches employed in mobile and wireless technologies. The security protocols in which the algorithms are employed are not described herein but can be found in each technology-specific chapter.

With the current trends toward a highly mobile workforce and wireless communications, the acquisition of mobile and wireless devices is growing at an ever-increasing rate. These devices offer productivity tools in a compact form and are quickly becoming a necessity in today's business environment. Mobile platforms are characterized by small physical size, limited storage and processing power, restricted user interface, and means for synchronizing data over short distances with other devices using radio signals or infrared.

Today, mobile platforms can send and receive electronic mail, access the Internet, manage appointments and contact information, exchange documents, deliver presentations, and access corporate data. Also, because of their relatively low cost, they are becoming ubiquitous within office environments, either provided by corporate information technology (IT) or purchased by the employees themselves as efficiency tools. However, several major issues loom over the use of mobile and wireless devices, including weak access control and unprotected wireless transmission. Such security weaknesses can provide new avenues for the introduction of viruses or other types of malicious code, as well as other forms of attack such as a man-in-the-middle attack. One of the major concerns in today's mobile e-business is security: service providers are changing their business models to exploit expanded access to their business-critical data and resources from mobile platforms. Such devices are the next vector for hackers and malware writers, who are now more interested in financial gain than earning some fame.

This chapter describes the security aspects of the most critical software components of mobile platforms, including communications stacks, OSs, Web services, digital rights management, and managed runtimes.

5.2   The New Risks

Mobile platforms increasingly retain corporate information, but unlike their desktop counterparts, they lie at the periphery of organizational controls and oversight. Limited computing power, memory, interfaces, and battery life impose constraints on the practicality of applying standard safeguards. The mobile platforms' small size and mobility also lead to greater exposure to theft or misuse in the field. Serious security concerns stem from the variety of ways in which a mobile platform can interact with other computing resources. These devices can inadvertently transfer malicious applications from one mobile platform onto another, or throughout the corporate network. Since mobile-platform-enabled, application-level malware cannot typically be blocked by corporate firewalls, a mobile platform may serve as a back-channel through which network vulnerabilities can be exploited.

In short, a mobile platform is exposed to multiple risks associated with external communications and interfaces over which corporate security officers exercise only limited control.

5.3   Elements of Mobile Security

Secure mobile software requires many components, among them software or hardware isolation functions and system integrity to ensure that misbehaving or malicious applications and users cannot affect other applications or users, system-level security to control and monitor the actions of users and applications on the platform, network-level security to protect other systems from outside attackers on the Internet, and transaction-level security to provide protection for business transactions on the Internet. Integrated cryptography support underlies a solid network- and transaction-level security implementation, and aids in providing system-level security. Finally, integration of the system security functions with system-provided applications, vendor-provided applications, and custom applications completes the picture.

System integrity means that unauthorized users and programs cannot bypass the software or hardware isolation functions that protect other users or programs, cannot obtain control in an authorized execution state, and cannot bypass the system-level security functions.

5.4 Communications Software Security

The best way to understand communications software security on mobile platforms is to understand the open system interconnection (OSI) model [1, 2]. The OSI model is an ISO standard for worldwide communications that defines a networking framework for implementing protocols in seven layers (see Figure 5.1). The OSI model divides the network into easily understood components that can be secured individually. Once each component is secured, the end-to-end security needs to be addressed.

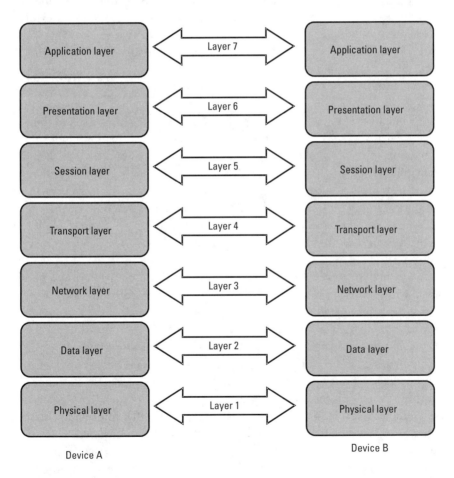

Device A

Device B

Figure 5.1 The OSI model.

In the OSI model, control is passed from one layer to the next one. Starting at the application layer, information is passed to lower layers until it reaches the bottom layer. Once the information reaches the physical medium, the information makes its way to the destination. When the information reaches the destination, it travels up each layer until it reaches the appropriate level for translation. An e-mail message, for example, starts at the application layer at the source device and makes its way down the stack, across the wire, up the stack to the destination device's application layer.

Within the source device, control is passed from one layer to the next. Data travels down the source device's hierarchy and then up the destination device's hierarchy. Figure 5.1 illustrates this flow of information. Notice that there is no way of skipping a layer and that the process is mirrored between communicating devices.

The physical layer communicates with the data link layer and the medium itself. Each layer is developed independently. This allows flexibility and allows development in one layer to progress without delays from other layers. As information passes through each layer, relevant information to that layer is attached—this process is commonly known as encapsulation. This encapsulation is how each layer can communicate with its relevant layer at the destination.

The physical layer (OSI layer 1) defines the physical properties of the network, such as voltage levels, cable types, and interface pins. Exploiting the physical requires some physical action, such as disrupting a power source or changing some interface pins.

The data link layer (OSI layer 2) transmits and receives packets of information reliably across a uniform physical network. Data link layer exploits include ARP cache poisoning, where an attacker alters the address resolution protocol (ARP) cache so that the wrong MAC address is associated with an IP address.

The network layer (OSI layer 3) routes data through various physical networks while traveling to a known host. Routers make decisions based on this layer's information, and routers base routing decisions on the IP. Layer 3 vulnerabilities include password buffer overflow.

The transport layer (OSI layer 4) ensures the reliable arrival of messages and provides error-checking mechanisms and data flow controls. One way the transport layer ensures that there is reliability and error checking is through the transport control protocol (TCP). Another protocol used at Layer 4 is the user datagram protocol (UDP). However, an attacker can gather information about a system using TCP and UDP. There are several ways in which TCP and UDP are used to infiltrate, deny services, or scan networks. The layer 4 attack of choice is the port-scanning attack, which is often an attacker's first probe of a network connection. Proper use of a firewall can prevent this type of attack.

The session layer (OSI layer 5) manages the setup and removal of the connection between two communicating end points. A connection is maintained while the two end points are communicating during a session. Layer 5 vulnerabilities include TCP session hijacking, when an attacker takes over a TCP session between two devices. Since most authentications occur only at the start of a TCP session, this allows an attacker to gain access to a device or launch a man-in-the-middle attack during a session. A common component of such an attack is to execute a DoS attack against one end point to stop it from responding. This attack can be against either the device or the network connection to force heavy packet loss. Solutions that provide some protection against layer 5 attacks include SSL and Internet protocol security (IPSEC).

The presentation layer (OSI layer 6) ensures that the information is acceptable to the application and session layers. ASCII and binary interpretations are presented to applications by layer 6 in presentations such as unicode. Layer 6 attacks include expressing malicious commands in unicode and requesting a Web server running on the device to execute those commands.

The application layer (OSI layer 7) defines standards for interaction at the user or application-program level. It is the highest layer of the protocol stack.

5.5 Mobile OS Security

A mobile OS is responsible for the administration of the platform, files, memory, and processes and is loaded directly after booting. The OS used on the mobile platform has a pivotal importance to ensure the security of the software running on the platform.

An attacker can install malware into the mobile platform in different ways: via an Internet connection, using a peripheral device such as a memory card, or by synchronizing the platform with any PC and installing malware [3]. The following sections describe various protection mechanisms that a mobile OS can take advantage of in order to provide additional protection to the mobile applications and services running on the platform.

5.5.1 Data Encryption

The mobile OS should provide the ability to encrypt data on the platform. This data protection mechanism should use encryption technology to protect all data, or only predefined sensitive data, on the mobile platform in the event it falls into unauthorized hands.

The mobile OS should also support automatic data encryption, which can take place automatically in the background without user intervention, in which case the only action the user must perform is to provide authorization to decrypt and access the protected data.

5.5.2 Internal Communication

The mobile OS should secure the internal communication on the mobile platform. If there is no secure path between the applications and the kernel, then the communication is vulnerable.

Protocols such as TLS can be used to provide sufficient integrity protection to the data communication between the application and the OS.

5.5.3 Memory and Application Separation

In order to protect security-critical applications, the mobile OS should have the ability to separate memory blocks and applications effectively from each other.

The OS should prevent each application from adjusting its priorities, terminating other applications, accessing their memory, and preventing the switchover into low power modes. One way to achieve the separation of applications is to have an effective distribution of permissions and rights.

The OS can thus protect the internal processes by means of a strict distribution of permissions in the lowest layers. The OS can protect the platform against several kinds of malicious programs by a systemwide separation of memory, access, and input/output rights for processes and applications. Here is one way to implement the verification of a secure state: The OS should not give malicious programs all user rights (as in many current systems).

The user for the first time should have the possibility to check if the mobile platform is in a secure state and if the OS is secure. This is not possible in most current mobile platforms. The OS's user interface should reserve an indicator on the screen that is permanently under its control. Since this indicator is under the sole control of the OS, it cannot be misused by a compromised platform. If the display indicates that the user is communicating with the secure OS, the control of the display and keyboard solely lies with the secure OS.

5.5.4 Access Control

The mobile OS should protect user credentials and should not allow them to be deactivated by the user. The OS should prevent a manipulated program from the possibility of acting with all user authorizations.

For use in a corporate network, the mobile platform should be administered by means of an access control list (ACL). By using this list, certain contents can be protected from being accessed by other device management servers so that the data can only be synchronized with an authorized server.

5.5.5 Code Signing

The OS's security may be augmented by a code-signing mechanism, with which the origin of programs and device drivers can be verified.

Since it is possible to port malware onto a mobile platform, some malware could later fake a signature. Fake dialogs are also possible because of malware and may be used to bypass signature verification. As some OSs support Active X and Java applications, these can be used to create fake dialogs. The user loads a Java applet or ActiveX control from a Web server, which is then executed on the mobile platform. The applet or control makes use of the owner's authorizations to gain access to the company's database and then copies data onto the mobile device. The applet or control then sends the data obtained back to a remote Internet server. In case of a Java applet, the sandbox restricts the applet's access to the hardware and software. However, the user may have granted the applet too many rights or an attacker may use a security gap in the Java virtual machine.

5.5.6 Direct Memory Access

Direct memory access (DMA) is supported by several mobile OSs through the support of several types of processors.

Device driver flaws can be more dangerous than other application vulnerabilities because device drivers are, in most cases, part of the OS itself, and subverting the critical software gives an attacker direct access to the kernel. Moreover, drivers that have DMA—such as USB drivers, CardBus drivers, graphics drivers, and sound drivers—could be used to overwrite system memory and exploit the system.

Also, since some device drivers can allocate, initialize, and free DMA-related resources, DMA services must have sufficient memory-page-protection checking and enforcement. DMA transfers between a device and memory must be based on enforceable access permissions, and protection violation must be detected.

5.5.7 Back Doors

There are tools that can be used by an attacker to bypass any security instruments in most mobile operating systems. With mobile maintenance programs (e.g., the Nokia Wintesla maintenance program [4]), several interventions in the mobile platform are possible, even when it is blocked. An attacker may obtain full access to many setup options of the device, unblock it with the knowledge gained, and gain full access to stored data. Any security claims for mobile platforms are thus reduced to absurdity if there is a back door.

5.5.8   Secure Boot

Secure boot checks the integrity of a mobile platform and the applications loaded into the memory. It has the ability to recognize virus activities, coding errors, or any malicious software core-configuration alterations.

When such integrity violations are detected, secure boot can limit the services offered by the platform. In a TPM-enabled platform, the trusted boot functions provide the ability to store in platform configuration registers (PCR), hashes of configuration information throughout the boot sequence.

Once booted, data (such as symmetric keys for encrypted files) can be *sealed* under a PCR. The sealed data can only be unsealed if the PCR has the same value as at the time of sealing. Thus, if an attempt is made to boot an alternative system, or a virus has back-doored the OS, the PCR value will not match, and the unseal will fail, thus protecting the data. The TPM's initialization and management functions allow the owner to turn functionality on and off, reset the TPM chip, and take ownership. This group of functions is somewhat complex, to provide strong separation of what can be done at BIOS (boot) time, and what can be done at normal runtime, so that sensitive operations can't be performed by malicious applications trying to compromise the platform's integrity.

5.5.9   Trusted Space

As described in Chapter 4, TCG hardware consists of two tamper-resistant modules, TPM and CRTM. Both of them will only be of use if an OS is used that supports them. Currently there are two OSs being developed that will support TCG hardware. Microsoft is developing a security technology that will be included in the Longhorn OS, and there are also initiatives to develop a Linux distribution that supports the TCG security modules [5].

Let's now look at how the TPM and CRTM can help build a trusted space where applications and data cannot be manipulated. The TPM hardware module can be regarded as an extended smart card on which secrets inside and outside of the TPM can be produced and stored [6, 7]. These secrets are symmetric and asymmetric keys that are used to ensure the trustworthiness of files, signing of data, and the authentication of third parties on the platform. Furthermore, hash values are examined to identify the trustworthy hardware and software components and are stored in PCRs. For a TPM to be active, its hardware must be switched on and software has to be activated.

For each component (BIOS, OS loader, and OS), a hash value is generated and transmitted to the TPM when the system is started. These values are stored in the *platform configuration register*. It is then examined to check whether the currently established hash values are identical with those stored on the TPM. If this is the case, the user can assume that the components and/or the data stored on them have not been manipulated, as otherwise the hash value would have changed and the system or the software would have informed the user. This way an authentication chain is established starting with the CRTM.

The OS can then build a trusted space for security-critical applications in which the applications are separated from each other, and any access from the outside into the *trusted space* is prevented. Uncertified programs, such as a virus or trojan horse, do not have access to the trusted space [7].

5.6 Web Services Security

5.6.1 Introduction

XML and Web services are clearly the foundation of a new generation of mobile applications. XML is a markup language derived from SGML, and it has garnered a large amount of support due to its ability to describe data. Because of the wide use of XML in emerging mobile Web services applications, this section will describe the various security aspects of Web services.

In contrast to client-server communications, Web services are designed to seamlessly connect resources above the network layer, thus enabling the concept of loosely coupled but tightly contracted applications. These applications enable easy direct access to valuable backend databases and application servers and, therefore, require the fine-grained control of granular security policies above the network layer.

In the context of Web services, security means that a message recipient is able to perform the following operations:

- Verify the integrity of a message;
- Receive a message confidentially;
- Determine the identity of the sender;
- Determine whether the sender is authorized to perform the operation requested by the message.

5.6.2 Transport Layer Security

XML-based Web services rely on IP and most typically hypertext transport protocol (HTTP) as a transport layer to connect applications and associated resources to one another. Because XML is a text-based data-encoding standard, it is human readable and easily deciphered when intercepted. Consequently, robust Web services security is built on a strong foundation of transport layer security so that messages cannot be intercepted and read in transit.

While transport layer security such as SSL or TLS cannot provide the more granular security functions, it is a minimum requirement for ensuring confidentiality of information during transport. Using server and client certificates during authentication helps protect against the known weaknesses associated with using source IP or domain name server information for access control or authentication.

5.6.3 Protection Against XML Denial of Service (X-DoS) Attacks

Whereas IP-based DoS attacks require coordination of thousands of clients to simultaneously swamp a server with requests, an XML DoS (X-DoS) attack can be launched with a single low-bandwidth message that is undetected by an IP firewall. In some instances, even unintentionally malformed content can create a service outage. Recently issued security advisory bulletins have detailed how vulnerabilities in several vendors' XML parsers can enable maliciously malformed XML documents to create a DoS by consuming CPU cycles and memory. Because these traditional security devices lack the ability to differentiate bad XML from good XML, it is simple to forward a malformed or corrupted XML document directly to backend resources for processing. Once the message passes beyond traditional network-level security systems, it can compromise backend server resources by erasing data, exporting sensitive information, or consuming resources with an infinite processing loop. Rather than a temporary bandwidth outage, this can result in serious and sometimes permanent disruption of service.

Several steps can be taken to protect against X-DoS. The first is to implement reasonableness constraints for all incoming messages. Configuration settings need to control message size, frequency, and connection duration [8–10].

5.6.4 Message Validation

Since XML is text-based and in many instances generated by humans, there is significant room for error in message creation. Whether data is malformed intentionally or unintentionally, it can consume valuable server processing cycles without warning, resulting in service degradation or complete outages. XML schema validation may be used to ensure that both incoming and outgoing messages are valid.

Message validation can reduce the risk of security vulnerabilites of unknown fields or protocol features that might otherwise compromise resources. Besides performing schema validation on all messages, mobile applications should also check messages for XML well-formedness (during parsing), improper identity or lack of resource references, protocol (e.g., SOAP) validity, and other message validity checks.

It is worth mentioning that, as with all other XML security functions, performing schema validation can be processing-intensive and challenging during times of peak usage [8–10].

5.6.5   Message Signature and Timestamping

Validating the identity of a Web service requestor provides authentication and prevents documents tampering during transmission. The IETF and W3C organizations have specified an XML digital signature standard that can provide this protection across an entire XML document or at the element level within a document. Signing and verifying every incoming and outgoing message can be processing-intensive.

Furthermore, when used with XML digital signatures, Web-service-based mobile commerce applications can have a cryptographically secure timestamp that enhances non-repudiation capabilities by being able to definitively prove at what time a given transaction took place [8–10].

5.6.6   Message Encryption

Transport layer security is designed to provide bulk encryption of the content being sent from one point to another. Once the content is delivered to the recipient, bulk decryption occurs and the entire message's content is visible. Because of XML's plain-text format and its multihop use model, this is a significant issue. The XML-based Web services model is different, as it is frequently multipoint in nature and requires that different portions of a message be selectively shared with recipients depending on their identities. As such, the XML encryption standard allows the encryption of individual XML documents and data fields within a document with different encryption keys. Unlike transport layer encryption, XML encryption is applied at a document or field level rather than at a packet level [8–10].

Since both XML encryption/decryption and XML processing (e.g., parsing, XPath selection, serialization, and other XML operations) are very resource-intensive, deploying both XML encryption and its companion, XML digital signature, can have a significant performance impact on high-transaction applications.

Because the underlying ciphers are the same, the keys and certificates used can also be the same as those used with other security applications and protocols such as SSL, potentially loaded from an existing PKI repository or key management system [8–10].

The use of common credentials (keys and certificates) can be an advantage but can also lead to vulnerabilities if the credentials are exposed by non–Web services applications.

5.6.7 Web Services Security-Related Standards

There are several security-related standards that are at different stages of development at the time of this writing. The most important ones are [8–10]:

- *XML Digital Signature:* The W3C/IETF group defined an XML digital signature specification.

- *XML encryption:* The W3C group defined an XML encryption specification.

- *XML Key Management:* The W3C group defined a specification to allow clients to obtain cryptographic key information (i.e., keys, certificates) and to perform key management such as initial registration and revocation.

- *OASIS Security Services Technical Committee (TC):* This group has been defining security authorization markup language (SAML), which is a framework for exchanging identification information; for example, a trusted third-party (such as a PKI CA or a network login server) could provide a signed set of assertions identifying my identity. SAML is the basis of the Liberty Alliance federated single sign-on architecture.

- *OASIS Access Control Markup Language TC:* This group has been defining Extensible Access Control markup language (XACML), which is a framework for defining a set of privileges required to perform an operation, including identity information and external factors (e.g., access policy).

- *OASIS Digital Signature Services TC:* This group has been defining an interface for a signature generation and verification service.

- *OASIS Web Services Security TC:* This group has built on the Web services' (WS) security (WSS) specification, which defines how to sign and encrypt a SOAP message in order to build a foundation for higher-level security services, such as policy integration and automatic interoperability.

- *Web Services Interoperability Organization:* This group has been defining a security profile to ensure basic interoperability among vendors.

5.6.8 Web Services Security (WSS) Specifications

At message-level security, security information travels along with the Web service message. In the SOAP layer WSS defines the use of XML encryption and XML digital signatures to secure SOAP messages. WSS profiles define the use of various security tokens, including X.509 certificates, SAML assertions, and username/password tokens in order to secure the messages.

As indicated earlier in this chapter, message layer security differs from transport layer security in that message layer security can be used to decouple message protection from message transport so that messages remain protected after transmission, regardless of how many hops they travel on.

WSS defines the binding of XML digital signatures, XML encryption, and username/password tokens to secure SOAP messages. The WSS specification defines a SOAP extension, which provides quality of protection through message integrity, message confidentiality, and message authentication. WSS mechanisms can be used to accommodate a wide variety of security models and encryption technologies.

The WSS specification provides an extensible mechanism for associating security tokens with SOAP messages. The WSS specification itself does not define the format of the various types of security tokens. Instead, a series of security token profile documents have either been published or are in the process of being published. Each profile document defines the use of a particular type of security token (e.g., X.509) to secure SOAP messages using digital signature or encryption.

Username token verification specifies a process for sending username tokens along with the message. The receiver can validate the identity of the sender by validating the digital signature provided by the sender. A digital signature internally refers to a security token (e.g., username token or an X.509 certificate token) to indicate the key used for signing. Sending these tokens with a message binds the identity of the tokens (and any other claims occurring in the security token) to the message [8–10].

5.7 Content Protection

5.7.1 Introduction

The sharing of media and entertainment via mobile platforms is becoming an increasingly popular pastime and one of the most widely used mobile services. Users download content to their mobile phones or receive information by MMS every day, thereby allowing content to be passed along from one to the other.

Content providers and mobile carriers are facing piracy issues similar to those caused by peer-to-peer networks on the Internet, and they are losing revenues since much of today's lower-value content is forwarded from one user to the next for free. As new smart phones and other mobile devices with color displays and richer audio capabilities penetrate the market, and as network capacities increase thanks to a growing number of WLAN hotspots, consumers are demanding access to higher-value content. Recognizing the revenue potential of these services, mobile carriers and content providers aim to fulfill these consumer demands, while still looking to protect their investments in high-value content.

Addressing the most critical dilemmas in the life cycle of premium content—intellectual property, integrity protection, security, and privacy—successful DRM solutions enable the operation of high-quality mobile services with secured revenues, while also allowing super distribution—the easy, secure forwarding of content from one person to another.

DRM solutions need to work across different devices, geographies, operators, and mobile terminals. They need to escort protected files wherever they go and enforce administrator-defined policies, including who can read what, which content can be duplicated or shared, and how long a user can view a file. Without a secure and interoperable DRM solution, the full potential of mobile media and entertainment delivery cannot be realized [11, 12]. A typical DRM conceptual architecture is shown in Figure 5.2.

5.7.2   The Open Mobile Alliance DRM Specifications

In late 2002, the Open Mobile Alliance (OMA) released the OMA DRM version 1.0 enabler [13], its first set of specifications based on a subset of the open digital rights language (ODRL) rights expression language. Designed to protect light media content (such as ringtones, wallpaper, Java games, video and audio clips, and screensavers), OMA's first DRM enabler includes three levels of protection and functionality: forward lock, combined delivery, and separate delivery, each level adding a layer of protection on top of the previous level.

The first level, forward lock, prevents the unauthorized transfer of content from one device to another. The intention is to prevent unauthorized

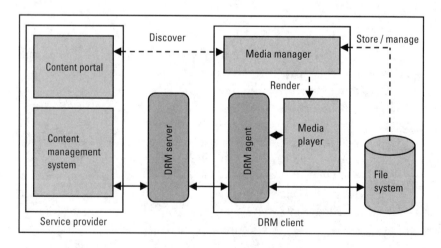

Figure 5.2 DRM conceptual architecture.

peer-to-peer distribution, or super distribution, of lower value content. Often applied to subscription-based services, such as news or sports, the plaintext content is packaged inside a DRM message that is delivered to the terminal. The device can play, display, or execute the content, but not forward the object.

The second level, combined delivery, also prevents forwarding, and it controls the content usage when rights definition is added to the first level. The DRM message contains two objects: the content and a rights object. The rights object, written into the content using OMA rights expression language, a mobile profile of ODRL, defines usage rules that govern the content. The rules include and support all kinds of business models, including preview and time- and usage-based constraints (e.g., a complimentary preview—the permission to play a tune only once, using the content only for a specific number of days, or an annual subscription with noninterfering price models). When applying the combined delivery mechanism, neither content nor the rights object can be forwarded from the target device.

The third level, called separate delivery, is the most advanced DRM mechanism because here the content is encrypted, thereby providing better protection for higher value content. Encrypted into DRM content format using symmetric encryption, the content is useless without a rights object and the symmetric content encryption key (CEK), which is delivered separately from the content. OMA requires that the CEK is delivered securely via wireless access protocol (WAP) push directly to the authorized mobile platform, where the DRM user agent uses it for content decryption.

An OMA DRM-compliant mobile platform securely stores the rights objects outside of the consumer's reach. Only the media player on that device has access to both encrypted content and the rights object including the CEK, in order to enable the consumption of the content by displaying or playing it.

Mobile users can download media and entertainment content and forward it to friends via MMS, but the recipients will not be able to use the content until they obtain their own CEK for content decryption. A *rights refresh* mechanism enables recipients of super-distributed content to contact the content provider to obtain rights to either preview or purchase the content they have received.

The DRM v.2.0 enabler specifications take advantage of expanded device capabilities and provide improved support for audio and video rendering, streaming content, and access to protected content using multiple devices. They have added security and trust certificates that allow more complex and rich forms of media content (i.e., premium content such as music tracks, video clips, and animated color screensavers and games) as well as improved support to preview and share content.

Security is enhanced by encrypting the rights object and the content encryption key, using the device's public key to bind them to the target device. Integrity protection for both content and the rights object reduces the risk of either being tampered with. The specifications also include mutual authentication between the device and the rights issuer (the content provider).

The OMA DRM v.2.0 implements a DRM that is defined as a mobile profile of the ODRL. ODRL is an expression language that addresses protection of sensitive information and purchased content that is in possession of the customer, the prevention of unauthorized use and distribution of content, and the avoidance of tampering with content, either during transmission or as a case of unauthorized reuse [11–13].

5.7.3 License-Driven DRM Standards

The following are the leading license-driven DRM standards for content protection of media content:

- *Content scrambling system:* A proprietary licensable encryption scheme that is used to encrypt the MPEG-2 payload on DVD video disks. Keys need to be obtained from the DVD Copy Control Association.

- *Content protection for prerecorded media:* A licensable copy protection method from the 4C Entity; this is a variant of CPRM for prerecorded media, descended from the content scrambling system. Content protection for prerecorded media is used on DVD audio disks.

- *Content protection for recordable media (CPRM):* A proposed renewable cryptographic method from the 4C Entity for protecting entertainment content when recorded on physical media.

- *Digital transmission content protection:* A proposed encryption mechanism for use on advanced digital interconnect joining consumer electronics and PCs, sponsored by the 5C entity. The main concern of digital transmission content protection is that unencrypted media transmitted over standardized high-speed digital interconnect such as IEEE 1394 is easily intercepted for piracy purposes.

5.7.4 Main Content Protection Organizations

The following organizations foster some aspects of defining content protection of media:

- *The 4C Entity:* A consortium of four computer technology companies (IBM, Intel, Matsushita, and Toshiba) that foster the production of, and licensing of intellectual property associated with content control.

- *The 5C Entity:* A consortium of five computer technology companies (IBM, Intel, Matsushita, and Toshiba, who are the 4C Entity, plus

Hitachi) that fosters the production of, and licensing of intellectual property associated with content control. The 5C entity emphasizes secure transmission (e.g., over domestic IEEE 1394 links), while the 4C Entity emphasizes secure storage;

- *Advanced Access Content System Licensing Authority:* The licensing authority that is developing the advanced access content system, a specification for managing content stored on the next generation of prerecorded and recorded optical media for use with PCs and consumer electronics devices;

- *Copy Protection Technical Working Group:* An industry consortium, supported by the Motion Picture Association of America, that proposes copy protection technology. They created the Broadcast Flag proposal;

- *Digital Content Protection LLC:* An organization created to license high-bandwidth digital content protection, a scheme for protecting video on DVI links.

- *The Digital Video Broadcasting Project:* The Digital Video Broadcasting Project is an industry consortium concerned with several aspects of digital television technology.

- *Moving Picture Experts Group (MPEG) Licensing Authority:* A one-stop shop for intellectual property licensing related to video and DRM. Implementers of modern media systems involving DRM are at risk of infringing dozens of patents. The effort and risk of researching and licensing DRM-related technologies piecemeal is huge, so organizations like to MPEG Licensing Authority to offer access to appropriate patent pools for specific technologies.

- *Smartright:* A consortium of mostly European companies that supports a smart–card-based copy protection system for digital home networks.

- *TV-Anytime Forum:* An organization of mostly European companies that seeks to develop specifications to enable audio-visual and other services based on mass-market high volume digital storage in consumer platforms—simply referred to as local storage.

5.8 Managed Runtime Security

5.8.1 Introduction

Managed runtime environments are considered an essential design element of all new mobile platforms by providing mobile application developers, operators, and end users an array of benefits, including

- A platform-independent programming environment that makes it easier (than native code) to move applications between platforms;

- A dynamic code-loading mechanism that makes it easier to extend platform capabilities with new applications and class libraries;

- A sandbox runtime environment that can prevent rogue programs from disrupting the platform;

- Garbage-collection memory management and incorrect-reference (pointer) protection that together attempt to eliminate a major source of exploitable programming.

Since Sun Microsystems' introduction of a small-Java solution and Java virtual machine for memory-constrained devices in 1999, the inherent protection from memory overruns provided in the Java architecture has made Java attractive. Wireless operators are highly sensitive to mobile platforms suspending operations without warning while a customer is operating the device and to the capability of proliferating network-damaging viruses. The favorable response to Java was also due to its inherent memory efficiency of bytecodes versus native code and to the open Java standardization efforts within the Java Community Process [14].

Currently, Java wireless client devices are based on the Java 2 platform, micro edition (J2ME), and specifically on the connected limited device configuration (CLDC) version 1.0 and the mobile information device profile (MIDP) version 1.0 [15].

The CLDC 1.0 provides the platform that is intended to serve as a common software layer for several types of mobile platforms, such as mobile phones and point-of-sale terminals, while the MIDP 1.0 addresses the specific needs of mobile phones. CLDC 1.0 addresses support for the Java language and virtual machine features, core libraries, input/output, networking, security, and internationalization. MIDP 1.0 addresses application models, user interfaces, persistent storage, networking, and timers [15, 16].

In the case of CLDC 1.0, there is no support for finalization, no support for the Java native interface, no support for user-defined class loaders, no reflection, no support for thread groups or daemon threads, no support for weak references, or advanced exception handling,

MIDP 2.0 and CLDC 1.1 [Java specification request (JSR)-118 and JSR-139, respectively] address many of the concerns of MIDP 1.0 and CLDC 1.0. CLDC 1.1 has added support for floating point as well as support for weak references. MIDP 2.0 provides significant improvements over MIDP 1.0. It provides enhanced networking support, enhanced user interface support, support for gaming, support for sound, and a security model that better aligns with the J2SE model of security [14–18].

Other major enhancements include support for optional packages, which are intended to supplement the functionality provided by CLDC 1.1 and MIDP 2.0 and eliminate the need for the use of original equipment manufacturer classes. Examples of optional packages include JSR-120 (wireless messaging API), JSR-135 (mobile media API), JSR-82 (Bluetooth), JSR-80 (USB), JSR-177 (security and trust APIs), JSR-172 (WSs) and more. The Java specification that ties all of these packages together is JSR-185—Java Technology for the Wireless Industry [15].

5.8.2 Java Security

Java security has the following two goals [14–18]:

- Provide the Java platform as a secure, ready-built platform on which to run Java-enabled applications in a secure fashion;
- Provide security tools and services implemented in the Java programming language that enable a wider range of security-sensitive applications (such as corporate applications).

The original security model provided by the Java platform is known as the sandbox model, which was defined to provide a restricted environment in which to run untrusted code obtained from the open network. The essence of the sandbox model is that local code is trusted to have full access to vital system resources, such as the file system, while downloaded remote code, such as an applet, is not trusted and can access only the limited resources provided inside the sandbox. The sandbox model was deployed through the Java development kit (JDK) and was generally adopted by applications built with JDK 1.0, including Java-enabled Web browsers.

Java security is enforced through a number of mechanisms. First, the language is designed to be type-safe and easy to use. The goal is to lessen subtle programming mistakes compared with other programming languages such as C or C++. Language features such as automatic memory management, garbage collection, and range checking on strings and arrays also help the programmer to write safe code. Second, compilers and bytecode verifiers ensure that only legitimate Java bytecodes are executed. The bytecode verifier, together with the Java virtual machine, guarantees language safety at run time. Moreover, a *classloader* defines a local name space, which can be used to ensure that an untrusted applet cannot interfere with the execution of other programs. Finally, access to crucial system resources is mediated by the Java virtual machine and is checked in advance by a *SecurityManager* class that restricts the actions of a piece of untrusted code to the bare minimum.

Furthermore, JDK 1.1 introduced the concept of a digitally signed applet, which is treated as if it is trusted local code when the signature key is recognized as trusted by the mobile platform that receives the applet. Signed applets, together with their signatures, are delivered in the Java archive format. In JDK 1.1, unsigned applets still run in the sandbox.

The Java 2 platform security architecture, shown in Figure 5.3, further provides the following capabilities:

- *Fine-grained access control:* Java 2 makes using fine-grained access control simpler and safer than JDK 1.1, where customization of the *SecurityManager* and *ClassLoader* classes is required.

- *Easily configurable security policy:* Java 2 makes configurable security policies simpler and easier to use than JDK 1.1.

- *Easily extensible access control structure:* Java 2 allows typed permissions, each representing an access to a platform resource, and automatic handling of all permissions, including yet-to-be-defined permissions, of the correct type. No new method in the *SecurityManager* class needs to be created in most cases:

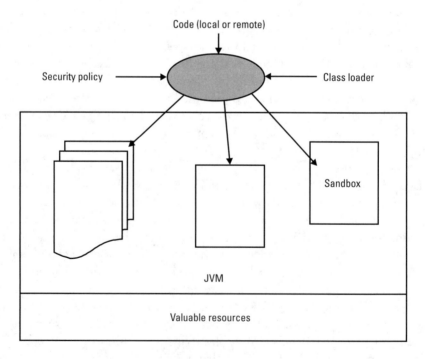

Figure 5.3 Security architecture of the Java 2 platform.

- *Security checks:* Java 2 provides an extension of security checks to all Java programs, including applications as well as applets.

- *Trusted code:* There is no longer a built-in concept that all local code is trusted. Local code is subjected to the same security controls as applets, although it is possible, if desired, to declare that the policy on local code (or remote code) be the most liberal, thus enabling such code to effectively run as totally trusted. The same principle applies to signed applets and any Java application.

5.8.3   Critical Java Security APIs

The following APIs extend the Java 2 security capabilities by providing additional security services that greatly enhance mobile platforms that make use of the Java runtime environment [19–21]:

- *Java cryptography extension:* The Java cryptography extension is a set of packages that provides a framework and implementations for encryption, key generation and key agreement, and message authentication code (MAC) algorithms. Support for encryption includes symmetric, asymmetric, block, and stream ciphers. The extension also supports secure streams and sealed objects.

- *Java authentication and authorization service:* The Java authentication and authorization service is a set of APIs that can be used for authentication and authorization of users to reliably and securely determine who is currently executing the Java code. In Java 2, permissions are granted based on code characteristics: where the code was coming from and whether it was digitally signed and, if so, by whom. With the integration of the Java authentication and authorization service into the Java 2 software development kit (SDK), the *java.security.Policy* API handles principal-based queries, and the default policy implementation supports principal-based grant entries. Thus, access control can now be based not just on what code is running, but also on who is running it.

- *Java secure socket extension:* The Java secure socket extension is a set of packages that enable secure Internet communications. It implements a Java technology version of SSL and TLS protocols. It includes functionality for data encryption, server authentication, message integrity, and (optional) client authentication.

- *Security and trust services API for J2ME; JSR 177:* The security and trust services API for J2ME extends the security features for the J2ME environment through the addition of cryptographic APIs, digital signature service, and user credential management. This specification also defines the methods to communicate to a smart card by leveraging the application protocol data unit (APDU) protocol and Java card remote method invocation (see section 4.2.1 for more details.)

5.8.4   The .NET Managed Runtime Environment

In 2000, Microsoft introduced .NET, a framework intended to accomplish many of the same tasks as Java. Much like the deployment of Java, the larger, enterprise-focused .NET was available prior to the smaller footprint Compact Framework .NET, which began to roll out gradually in the 2002 releases of Microsoft Windows CE.

.NET framework has a number of differences when compared to other runtime environments, including Java. First, .NET supports a number of different languages including C, C++, C#, Visual Basic, and JavaScript. Using Microsoft's Visual Studio .NET, programs written in these languages are compiled into a common intermediate language representation that executes within the common language runtime (CLR) environment.

The two major components of the .NET Framework include [20, 21]:

- *The CLR:* The CLR is the execution engine for .NET Framework-based applications. Its functions include code management (loading and executing code), managing memory for applications and objects, enforcment security rules regarding what resources may be accessed by the code, and interoperation between .NET Framework-based code and preexisting complement object model (COM) objects.
- *Class libraries:* They provide functionality for tasks such as user interface design, threading, security management, and network communications. The three most significant groups of classes are ADO.NET for data manipulation, ASP.NET for building Web applications and XML Web services, and Windows Forms for building Windows-based smart-client applications.

These components of the .NET Framework were designed to facilitate the inclusion of a broad range of security options. The security features of the .NET Framework include [20, 21]

- *Role-based security:* Role-based security provides a unified model for authenticating and authorizing users based on identity and roles. Authentication involves examining user credentials. Once authentication is established, application code can determine what role the user has and what operations he or she is allowed to perform. The .NET

Framework provides support for common authentication protocols, including Basic, Digest, NTLM, Kerberos, and SSL/TLS client certificates.

- *Evidence-based and code access security:* The .NET Framework provides a unified mechanism to define the resources that can be accessed by certain types of code with evidence-based and code access security (e.g., code residing in a particular directory, code coming from the Internet or an intranet, code with a certain hash value, or code signed with a certain key). This layer of security provides a granular control over the devices running in any specific domain, protecting them from potentially malicious code.

- *Cryptography:* The .NET Framework cryptography includes functions for encryption, digital signatures, hashing, and random number generation. Algorithms supported by the .NET Framework include symmetric encryption (DES, Triple-DES, RC2); asymmetric encryption (RSA, DSA); the XML digital signature specification; and hashes (MD5, SHA1).

5.9 Access Control

5.9.1 Authentication

Since the use of mobile platforms within the enterprise is expanding rapidly, they pose new risks to an organization's security, not only from the sensitive information held and the organizational networks accessible by them, but also from their propensity to become physically separated (e.g., stolen) from the user. Adequate user authentication is the first line of defense against unauthorized use of a lost or stolen handheld device. Multiple modes of authentication increase the work factor needed to compromise a device.

In order to reduce or eliminate authentication risks, mobile platforms must have the means to express, monitor, and enforce organizational security policies effectively, particularly over external communications and interfaces. Policies should not only restrict and filter external communications, but also constrain user privileges on mobile devices. When implemented correctly, security policy management mechanisms can be applied to govern user behavior automatically [22].

Also, mobile platforms must protect user credentials. For example, once the original password is entered, it is then discarded and only the protected form of the password remains. To improve their resistance to attack, a randomly generated *salt* value may be concatenated with the original password before cryptographic transformation. The addition makes it harder for the attacker to assemble precomputed dictionaries of common passwords [23].

Graphical login can also be used. Graphical login refers to a class of authentication mechanisms that rely on the creation of graphical images to produce a password value. Graphical login is somewhat similar to visual login and shares many of the same attributes [24–26, 28].

Fingerprint verification can be a convenient method of establishing a user's identity. While biometric systems reduce the need for password management, enrollment and verification times can be an issue. Usually several biometric images are needed to create a template during enrollment. That process can be lengthened by the occurrence of a poor image, which also negatively affects verification times [23].

Biometric systems have variability when measuring human characteristics or behavior. The false rejection ratio indicates the percentage of authentication sequences that result in failure to authenticate a valid user. The false acceptance ratio indicates the ratio of authentication sequences that result in the authentication of an attacker. Ideally, both ratios should be as low as possible. Biometric systems usually have the means to set threshold levels tighter or looser to increase or decrease the level of security as required. A tight threshold setting reduces the likelihood for false acceptance errors, but increases the likelihood of false rejection errors. A loose threshold setting has the reverse effect [27, 28].

5.9.2   Policy Enforcement

Policy enforcement occurs on a mobile platform to ensure that the user adheres to the granted policies (e.g., the specific enterprise policies). The various policy enforcement mechanisms implemented should supplement, rather than replace, existing OS security mechanisms. The policy enforcement engine comprises two logical components: a policy manager and a policy enforcer.

The components of the mobile platform policy manager are responsible for obtaining certificates, validating them, extracting the policy entries from the policy certificate, and passing the entries to the policy enforcer.

The policy enforcement process begins with the enrollment of a user. During enrollment, a security officer generates an identity certificate and policy certificate for the user. The certificates are then issued to the user (possibly stored on a smart card). The enrollment may involve interaction with an X.509 CA and a policy CA to obtain the identity and the policy certificates for the user [29].

References

[1]   Tipton, H.F., and M. Krause, *Information Security Management Handbook,* 4th ed., Boca Raton, FL: Auerbach Publications, 1999.

[2]   Tanenbaum, A., and M. Van Steen, *Distributed Systems Principles and Paradigms,* Upper Saddle River, NJ: Prentice Hall, 2002.

[3]   Nicholls, R., and P. Lekkas, *Wireless Security: Models, Threats and Solutions,* New York: McGraw-Hill, 2001.

[4]   WinTesla v.5.31 Nokia Service Software for Windows, http://ucables.com/nokia/service/wintesla.html.

[5]   MacDonald, R., et al., "Bear: An Open-Source Virtual Secure Coprocessor Based on TCPA," http://www.cs.dartmouth.edu/~sws/papers/msmw03.pdf.

[6]   Trusted Computing Group, http://www.trustedcomputing.org/home.

[7]   Pearson, S., et al., *Trusted Computing Platforms: TCPA Technology in Context,* Upper Saddle River, NJ: Prentice Hall, 2002.

[8]   Dournaee, B., *XML Security,* New York: McGraw-Hill, 2002.

[9]   O'Neill, M., *Web Services Security,* New York: McGraw-Hill Osborne Media.

[10]   Rosenberg, J., and D. Remy, *Securing Web Services with WS-Security: Demystifying WS-Security, WS-Policy, SAML, XML Signature, and XML Encryption,* Indianapolis, IN: SAMS, 2004.

[11]   Rosenblatt, B., B. Trippe, and S. Mooney, *Digital Rights Management: Business and Technology,* M&T Books.

[12]   "Digital Rights Management for Interoperable Mobile Services Managing Content in the Next Few Years Is Only Going to Get More Complicated. Luckily Standards Are on the Way," http://wireless.sys-con.com/read/46640.htm?CFID=120676&CFTOKEN=B24BF A2C-554B-139C-DDB1B683C9E3550D.

[13]   OMA DRM, http://www.openmobilealliance.org/tech/docs.

[14]   Gong, L., *Inside Java 2 Security,* Sun Microsystems.

[15]   McGraw, G., and E. Felten, *Securing Java,* 2nd ed., New York: Wiley Computing Publishing, 1999.

[16]   Aissi, S., "Runtime Environment Security Models," *Intel Technical Journal,* Vol. 7, No. 1, February 19, 2003.

[17]   Drews, P., et al., "Managed Runtime Environments for Next-Generation Mobile Devices," *Intel Technical Journal,* Vol. 7, No. 1, February 19, 2003.

[18]   Comp, L., and T. Dobbing, "Runtime Abstractions in the Wireless and Handheld Space," *Intel Technical Journal,* Vol. 7, No. 1, February 19, 2003.

[19]   Mobile Information Device Profile, JSR 37, JSR 118, http://java.sun.com/products/midp/index.jsp.

[20]   LaMacchia, B., and S. Lang, *.NET Framework Security,* Boston, MA: Addison Wesley, 2002.

[21]   Freeman, A., and A. Jones, *Programming .NET Security,* Sebastopol, CA: O'Reilly, June 2003.

[22]   Norton, P., and M. Stockman, *Network Security Fundamentals,* Indianapolis, IN: SAMS, 2000.

[23]   McClure, S., J. Scrambray, and G. Kurtz, *Hacking Exposed,* 3rd Ed., Berkeley: Osborne/McGraw-Hill, 2001.

[24]   Microsoft, Let Me In: Pocket PC User Interface Password Redirect Sample, Microsoft Knowledge Base Article 314989, July 2003, http://support.microsoft.com/default.aspx? scid=kb;en-us;314989.

[25]   Northcutt, S., and J. Novak, *Network Intrusion Detection: An Analyst's Handbook,* 2nd ed., Indianapolis, IN: New Riders, 2000.

[26]   Jansen, W., "Authenticating Users on Handheld Devices," *Proceedings of the Canadian Information Technology Security Symposium,* Ottawa, ON, Canada, May 12–15, 2002.

[27]   Boertien, N., and E. Middelkoop, Authentication in Mobile Applications, CMG, Telematica Instituut, the Netherlands, January 2002, https://doc.telin.nl/dscgi/ds.py/ Get/File-23314/VH_authenticatie.pdf.

[28]   Jermyn, I., et al., "The Design and Analysis of Graphical Passwords," *8th USENIX Security Symposium,* Washington, DC, August 23–26, 1999.

[29]   Jansen, W., et al., "Security Policy Management for Handheld Devices," *The 2003 International Conference on Security and Management (SAM03),* June 2003.

6

Security Certification and Evaluation

6.1   Introduction

The goal of mobile security is the protection of information residing on mobile platforms from unauthorized disclosure, modification, or loss of use by countering threats to that information arising from human or systems-generated activities, malicious or otherwise. Countering threats to mobile platforms and mitigating risk helps to protect the confidentiality and integrity of information and ensure its availability.

Mobile platform OEMs, wireless operators, and other key entities in the mobile ecosystem need to have enough conviction that the security controls in the mobile platforms for their specific uses are effective in achieving the desired level of protection. In some circumstances, software cannot be used unless it has undergone a specific evaluation by an accredited entity. Users of mobile platforms not only need to have confidence in their security features, but they also want to be able to compare various products to understand their capabilities and limitations.

Security certification of a mobile platform (or a subset of its security features) is the comprehensive evaluation of its technical and nontechnical security features, along with other safeguards, to establish the extent to which a particular design and implementation meet a specified set of security requirements. Certification schemes can enable users of mobile platforms to obtain an impartial assessment of such products by an independent entity. This impartial assessment, or security evaluation, includes an analysis of the product and the testing of the product for conformance to a set of security requirements.

It is critical that security evaluations of mobile platforms are carried out in accordance with recognized standards and procedures. The use of standard security evaluation criteria and security evaluation methodology contributes to the repeatability and objectivity of the results.

Before it can be determined that a mobile platform is secure, it is necessary to determine the exact security requirements. Standards for identifying and defining security requirements such as the Common Criteria (CC) [1], standardized as the International Standards Organization (ISO)/International Electrotechnical Commission (IEC) 15408, can help achieve that goal. Some of the security certification schemes include specifications that are general and others that are focused on a specialized area—such as the National Institute for Standards and Technology (NIST) Federal Information Processing Standard (FIPS)–140 [2] specification for cryptographic equipment—and which provide benchmarks for implementing specialized security technologies.

To increase the level of confidence in such security evaluations, the final evaluation results can be reviewed by an independent party. An independent party can provide confirmation that the security evaluation has been conducted in accordance with the provisions of the scheme and that the results of the testing entity are consistent with the facts presented in the evaluation. This review can promote consistency of security evaluations and comparability of results for all evaluations conducted within the scheme. The impartial evaluation, the independent validation of evaluation results, and the documentation resulting from these processes provide valuable information about the security capability of mobile platforms to mobile OEMs, wireless operators, consumers, and all other entities involved in the mobile ecosystem.

This chapter provides an overview of several certification schemes that are the culmination of decades of work to identify security requirements and evaluating products to see if they meet the intended security requirements. Many of the schemes presented in this chapter are voluntary, such as the GSM Association's Security Accreditation Scheme [3]; some are required, such as the Cardholder Information Security Program [4]; and other schemes are provided as guidelines, such as the Organization of Electronic Cooperation and Development (OECD) Guidelines on the Protection of Privacy [5]. Although many of the certification schemes described in this chapter are not specific to mobile and wireless platforms, many of the security requirements that they target are applicable to such platforms. Furthermore, with mobile platforms becoming more and more open and computationally capable, the line between the security requirements for mobile platforms and for general computing devices (e.g., desktop PCs) is getting very thin. Therefore, the certification schemes, audit processes, and privacy guidelines discussed in this chapter are either used today or can be used in the future for mobile platforms.

6.2   Security Certification Schemes

6.2.1   The Common Criteria for Information Technology Security Evaluation

6.2.1.1   The CC

The Common Criteria for Information Technology Security Evaluation [1], simply known as CC, is the International Standard ISO/IEC 15408:1999 and includes requirements that list various kinds of security functions that products may include and various methods of assuring that a product is secure. The CC also has a related document, called Common Evaluation Methodology (CEM), for guiding evaluators on how to apply the CC when performing formal evaluations. Two documents are typically created from the CC:

- A protection profile (PP) is a document that defines the desired security properties of a product and lists the user security requirements, described using the CC definitions. There are several PPs, such as for smart cards, mobile code, and OSs. A PP can be used for similar products.
- A security target (ST) is a document that defines the product's security functions and can meet the requirements of one or more PPs.

The process of defining a PP or ST involves the following generic steps: identifying the security environment, deriving the security objectives for the product or product type, and finally selecting the security requirements that meet those objectives.

While the evaluation of a PP ensures that it meets the documentation rules, the evaluation of a ST ensures that the actual product, a target of evaluation (TOE), meets the security functional and assurance requirements. Functional requirements define what the product must do, and assurance requirements are measures to establish confidence that the security objectives are met. These evaluations are not proofs, but rather a measure that can help gain confidence in the requirements or product. In order to minimize subjectivity in the evaluation, the CC defines a set of assurance requirements called evaluation assurance levels (EALs), which range from 1 to 7. Several countries have signed mutual recognition agreements in order to accept evaluations done by accredited entities in other countries, and in most cases assurance measures of EAL-4 or less are required.

6.2.2   The Common Criteria Evaluation and Validation Scheme

The National Information Assurance Partnership (NIAP) created the Common Criteria Evaluation and Validation Scheme (CCEVS) [6] in order to establish a national program for the evaluation of information technology products for conformance to the CC. This program, which is currently managed by the NIST and the National Security Agency (NSA), approves participation of security testing laboratories in the scheme, provides technical guidance to security evaluation laboratories, validates the evaluation results for conformance to the CC, and works with other nations for the recognition of security evaluations.

Since this program also approves commercial testing laboratories based on the accreditation provided by NIST's National Voluntary Laboratory Accreditation Program, this accreditation is the main requirements for becoming a Common Criteria Testing Laboratory (CCTL). Several PPs have been evaluated and certified in accordance with the provisions of the CCEVS by accredited CCTLs in several countries participating in the CC program. Some of the validated PPs include key recovery, antivirus, biometrics, certificate management, tokens, firewalls, OSs, PKI/key management infrastructure (KMI), smart cards, and TCG. For example, the TCG PPs [7] describe the security requirements, threats, objectives, and EAL for the TPM, which provides security primitives such as digital signatures, random number generation, protected storage, and binding information to the TPM (see Chapter 4 for more details). Based on these TCG PPs, a TPM or mobile-platform OEM can create a ST that describes their evaluated TOE (i.e., TPM or mobile platform) and how these requirements can be met, and have this independently verified by a CCTL.

6.2.3   FIPS and the Cryptographic Module Validation Program

The FIPS 140-1, Security Requirements for Cryptographic Modules, and its successor FIPS 140-2 [2], are U.S. government standards that provide a benchmark for implementing cryptographic components. The FIPS standards specify best practices for the design and development of cryptographic algorithms, key management, and interaction with an OS. The FIPS evaluation process is administered by the NIST Cryptographic Module Validation (CMV) program, which allows encryption product vendors to demonstrate the extent to which their products comply with the standards.

Although some U.S. government agencies and commercial enterprises purchase only encryption products evaluated based on FIPS 140-1 or FIPS 140-2, the security community values more and more products that have completed this evaluation because it involves a clear set of requirements validated by an independent third party.

For software, the FIPS 140-1 or FIPS 140-2 evaluation process can include cryptographic services providers (CSPs) for encryption and decryption, public key certificate servers, Internet browsers (they typically contain certificates), TLS/SSL security providers, authentication and authorization modules, virtual private network (VPN) clients, and secure/multipurpose Internet mail extensions (S/MIME) email encryption protocols. For hardware, the FIPS 140-1 or FIPS 140-2 evaluation processes can include cryptographic modules, such TPMs and smart cards.

The FIPS 140-1 verification process is based on roughly six steps, as shown in Figure 6.1:

1. The vendor selects a test lab and submits the module for testing.
2. NIST/CSE issue testing and implementation guidance. The test lab submits questions for guidance and clarification and tests for conformance.
3. The module's test reports are published.
4. The test reports go to NIST/Communication Security Establishment (CSE) (in Canada) for analysis.
5. NIST/CSE issue a validation certificate.
6. NIST publishes a list of validated modules.

6.2.4   United Kingdom IT Security Evaluation and Certification

The purpose of the UK Information Technology Security Evaluation and Certification (ITsec) scheme [8] is to perform an independent security evaluation of

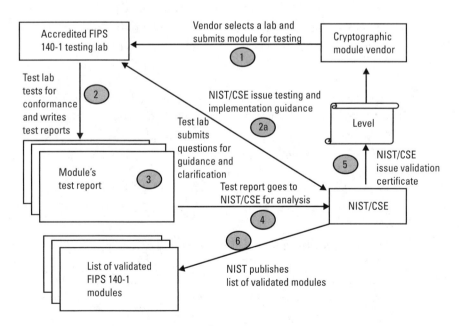

Figure 6.1 FIPS 140-1 verification process.

products according to standardized criteria using a formal method. The main goals of this exercise are to define a level of security for the product and to identify vulnerabilities. Although it is gaining fast adoption, the UK ITsec standard is less widely recognized that the CC, which is an ISO standard (ISO15408).

The ITsec evaluation criteria define seven levels of confidencelike or qualitylike marks known as assurance levels (ALs). These levels are labeled $E0$ through $E6$ and are defined as [8]:

- $E0$: Inadequate assurance is found.

- $E1$: Documentation gives guidance on TOE security. Security-enforcing functions are tested by evaluator or developer. TOE is to be uniquely identified and to have delivery, configuration, start-up and operational documentation, and secure distribution methods.

- $E2$: An informal detailed design and test documentation are produced. Architecture shows the separation of the TOE into security-enforcing and other components. Penetration testing searches for errors. Configuration control is assessed. Audit trail output is required during start-up and operation.

- $E3$: Source code or hardware drawings are to be produced. Correspondence must be shown between source code and detailed design. Acceptance procedures must be used. Implementation languages should be to recognized standards. Retesting must occur after the correction of errors.

- $E4$: Formal model of security and semiformal specification of security enforcing functions, architecture, and detailed design are to be produced. Testing must be shown to be sufficient. TOE and tools are under configuration control with changes audited and compiler options documented. TOE is to retain security on restart after failure.

- $E5$: Architectural design explains the interrelationship between security-enforcing components. Information on the integration process and run time libraries are to be produced. Configuration control is independent of developer. Configured items are identified as security enforcing or security relevant, with support for variable relationships between them.

- $E6$: Formal description of architecture and security enforcing functions are to be produced. Correspondence is shown from formal specification of security-enforcing functions through to source code and tests. Different TOE configurations are defined in terms of the formal architectural design. All tools are subject to configuration control.

For any claimed AL, certificates can be issued by the U.K. ITsec for products meeting the requirements, and these certificates are recognized in many countries. For example, certificates that are issued by the United Kingdom, France, and Germany are formally recognized by each of these countries, as well as by Finland, Greece, Italy, the Netherlands, Norway, Spain, Sweden, and Switzerland. Also, certificates issued by the UK, Australia, and New Zealand are recognized by each country.

6.2.5 TEMPEST

Telecommunications Electronics Material Protected from Emanating Spurious Transmissions (TEMPEST) [9] is defined as the study of the emission of unintentional protectively marked data from a product. The need for such a study results from the possible interception of unintentional radiation from an electronic product by a receiving device some distance away. The analysis of such radiation can reveal compromising emanations.

The UK national standards for TEMPEST testing and control provide a standard for protectively marked data. The standards that the UK requires for the suppression of TEMPEST signals carry a protective marking. TEMPEST certification in the UK is carried out by a test facility accredited by the Communications Electronics Security Group (CESG). The test results from these facilities are endorsed by CESG against the TEMPEST standards. Once endorsed by CESG, and a certificate is issued to the relevant standard, the product can be entered into a North Atlantic Treaty Organization (NATO) list of recommended products.

6.2.6 The GSM Association's Smart Card Security Accreditation Scheme

The smart card security accreditation scheme (SAS) [3] is a voluntary scheme that provides accreditation of smart card suppliers based on their ability to meet a defined set of security criteria and a security audit that covers security policy and procedures, information security, personnel security, physical security, IT security, as well as logistics and production management. The SAS's goal is to ensure that SIM suppliers and manufacturers are graded in accordance with specific standards and criteria.

The GSM Association's SAS accreditation scheme can award security certification for a period of two years on successful completion of an audit of the SIM supplier site by an approved security auditing company. Based on the audit results, a certification body makes the final decision whether or not the supplier can be awarded SAS certification.

6.2.7 Cardholder Information Security Program

Visa USA established the Cardholder Information Security Program (CISP) [4] to ensure account information safety and security assurance to their customers, who use their bankcard at points of sale and over the Internet. This program is intended to protect cardholder data, wherever it resides, to ensure that high standards of information security are maintained. With the emergence of mCommerce applications, a mobile platform will store, process, and transmit all or some part of a cardholder's data. Therefore, compliance to standards such as CISP will become more and more prevalent.

In order to achieve compliance with CISP, service providers have to meet the requirements of the payment card industry (PCI) Data Security Standard, which offers a single approach to safeguarding sensitive data for all card brands. This Standard is a result of collaboration between Visa and MasterCard, and is now endorsed by other credit card companies. The PCI Data Security Standard consists of the following 12 basic requirements:

1. Install and maintain a firewall configuration to protect data;
2. Do not use vendor-supplied defaults for system passwords and other security parameters;
3. Protect stored data;
4. Encrypt transmission of cardholder data and sensitive information across public networks;
5. Use and regularly update antivirus software;
6. Develop and maintain secure systems and applications;
7. Restrict access to data by business need-to-know;
8. Assign a unique ID to each person with computer access;
9. Restrict physical access to cardholder data;
10. Track and monitor all access to network resources and cardholder data;
11. Regularly test security systems and processes;
12. Maintain a policy that addresses information security.

6.3 Privacy Aspects

Software and hardware security components on mobile and wireless platforms must be designed and implemented with privacy in mind and adhere to the letter and spirit of all relevant privacy guidelines, laws, and regulations, such as the

OECD Guidelines, the Fair Information Practices, and, the European Union Data Protection Directive (95/46/EC).

6.3.1 The OECD Guidelines

The OECD issued a set of voluntary guidelines concerning the privacy of personal records [10], which are currently the basis of many governmental privacy rules, international agreements, national laws, and self-regulatory policies.

The OECD Guidelines have the following eight principles:

1. *Collection limitation:* This principle states that the collection of personal data must be limited and obtained by lawful and fair means and with the knowledge or consent of the data subject.

2. *Data quality:* This principle states that personal data should be relevant to the purposes for which they are to be used.

3. *Purpose specification:* This principle states that the purposes for which personal data are collected should be specified not later than at the time of data collection.

4. *Use limitation principle:* This principle states that personal data should not be disclosed, made available, or otherwise used.

5. *Security safeguards:* This principle states that personal data should be protected by reasonable security safeguards against such risks as loss or unauthorized access, destruction, use, modification, or disclosure of data.

6. *Openness:* This principle states that there should be a general policy of openness about developments, practices, and policies with respect to personal data.

7. *Individual participation:* This principle states that an individual should have the right to know whether or not a data controller has data relating to him and to challenge data relating to him.

8. *Accountability:* This principle states that a data controller should be accountable for complying with measures that give effect to the eight OECD principles.

6.3.2 The Privacy Protection Study Commission Fair Information Practices

The Privacy Protection Study Commission (PPSC) released a report entitled Personal Privacy in an Information Society, which includes recommendations on information practices to protect the privacy of industry-specific records [11]. Although the original report included a set of voluntary Fair Information Principles (FIP) for employers collecting personal data for hiring purposes, the FIP principles remain a useful guide for privacy in general. The PPSC-FIP includes principles that define disclosure of personal data, individual access to that data, data collection, authorizing the collection of personal data, handling medical records, the use of investigative organizations, and handling security records during conviction.

6.3.3   The European Union Data Protection Directive

The Data Protection Directive 95/46/EC [12] for Personal Data that was adopted by the Council of Ministers of the European Union (EU) granted individual persons a number of important rights, including the right of access to personal data, the right to know where the data originated, the right to have inaccurate data rectified, a right of recourse in the event of unlawful processing, and the right to withhold permission to use data in certain circumstances.

The main 95/46/EC principle is that personal data should not be processed at all, except when certain conditions are met. These conditions fall into three categories: transparency, legitimate purpose, and proportionality.

For transparency, the individual person (data subject) should have the right to be informed when his personal data is being processed. Data may be processed only under some specified circumstances, and the data subject has the right to access all data processed about him and to demand the rectification, deletion, or blocking of data that is incomplete, inaccurate, or isn't being processed in compliance with the data protection rules.

For the legitimate purpose category, personal data can only be processed for specified, explicit, and legitimate purposes and may not be processed further in a way incompatible with those purposes.

For the proportionality category, personal data can be processed only when it is adequate, relevant, and not excessive in relation to the purposes for which they are collected or further processed. When sensitive data is being processed, extra restrictions apply.

References

[1]   The Common Criteria for Information Technology Security Evaluation, http://www.commoncriteriaportal.org.

[2]   NIST FIPS-140, http://csrc.nist.gov/publications/fips.

[3]   GSM Association's Security Accreditation Scheme, http://www.gsmworld.com.

[4]   Cardholder Information Security Program, http://usa.visa.com.

[5]   OECD Guidelines on the Protection of Privacy, http://www.oecd.org.

[6]   Common Criteria Evaluation and Validation Scheme, http://niap.nist.gov/cc-scheme/defining-ccevs.html.

[7]   Trusted Computing Group, http://www.trustedcomputinggroup.org.

[8]   Communications Electronics Security Group, United Kingdom Information Technology Security Evaluation and Certification Scheme (ITsec), http://www.cesg.gov.uk.

[9]   Communications Electronics Security Group, TEMPEST, http://www.cesg.gov.uk.

[10] Organization for Economic Cooperation and Development, http://www.oecd.org.

[11] The Privacy Protection Study Commission Fair Information Practices (PPSC-FIP), http://www.fas.org/.

[12] The European Union Data Protection Directive (95/46/EC), http://ue.eu.int.

7

Higher Layer Security

Information is a very important asset, so when data is accessed over an open network such as the Internet, it is crucial for it to be accessible only to authorized parties. Security can be inferred at every level of the OSI stack. This chapter discusses end-to-end security solutions in layers above IP.

The requested security services differ according to the application executed. Home baking and private network access are some of the most frequent transactions and need peer authentication as well as data confidentiality and integrity. When accessing the enterprise private network, for example, mutual authentication should be inferred so that only authorized users can enter the network and so that rogue servers cannot be mounted to perform man-in-the-middle attacks. Confidential data must obviously remain concealed and integrity protected. In addition to these services, digital signature capabilities are useful for mobile commerce so that a user cannot deny executing a transaction and so that she can be sure she'll be debited for the correct amount for the product of her choice. Another service still frequently not supported is anonymity, a service whose need emerged recently to protect user privacy for certain applications, such as e-voting.

The main standards deployed for the security of client server applications are SSL [1] and TLS [2]. SSL was the first standard to be defined, designed by Netscape engineers. TLS is SSL's evolution, and it is today's de facto standard. TLS version 1 was published in 1999 as an IETF request for comment (RFC); extensions to the original version are under discussion.

Other standards were designed for specific purposes. Wireless transport layer security (WTLS) is used by the wireless access protocol (WAP); it's an adaptation of TLS optimized for mobile terminals with low computational capabilities. HTTP over SSL or TLS (HTTPS) is the secured version of the

HTTP format. Secure electronic transactions (SET) were defined by MasterCard for electronic payment. Secure shell (SSH) provides security when logging into another computer over a network, executing commands in a remote machine, and transferring files between machines.

7.1 SSL and TLS Protocols

The development of the SSL protocol has already ended, with version 3 being the last enhancement made. SSL is still used to secure legacy applications, but TLS is being deployed more and more. In this section, we will describe TLS 1.0 in detail and compare it to SSL v3. We will then mention the enhancements drafted for approval in TLS version 1.1, and we will point to initiatives for use of shared key TLS.

7.1.1 The TLS Protocol

The TLS protocol allows client/server applications communicating over the Internet to prevent eavesdropping, tampering, or message forgery. TLS must be layered on top of a reliable transport protocol, as is TCP. TLS is application protocol independent, meaning that any higher level protocol can be layered on top of it.

The TLS standard defines four record protocols: the handshake protocol, the alert protocol, the change cipher spec protocol, and the application data protocol. The handshake protocol consists of a suite of three subprotocols used for peers to agree upon security parameters to authenticate themselves and calculate keys for data encryption and integrity protection. The change cipher spec protocol notifies the receiving party that subsequent records will be protected under the newly negotiated algorithms and keys. Alert messages convey information on alert conditions; actions will be taken depending on the type of alert. The application data protocol transparently transports upper layer messages. We will concentrate on the handshake protocol, which is the first protocol to be executed and which defines security parameters for the communication.

The handshake protocol, depicted in Figure 7.1, is initiated by the client through a hello message. ClientHello and ServerHello messages allow the peers to agree on which algorithms to use, to exchange random values, and eventually to resume a session temporarily interrupted due to transmission failure. TLS allows calculation of shared keys based on public key cryptographic algorithms applied to data exchanged between client and server. Before public key cryptographic algorithms are executed, peer certificates should be verified. TLS specifies key derivation, starting from the premaster secret generation to encryption and integrity key refresh. Finally, Finished messages allow checking that the peer

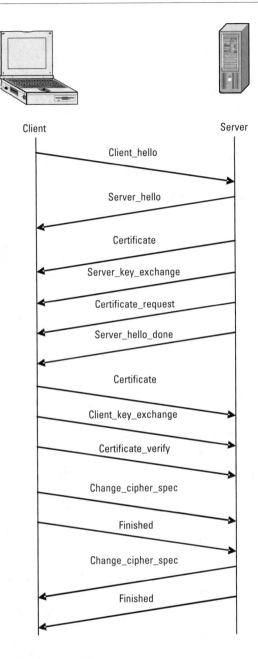

Figure 7.1 TLS full handshake protocol.

has calculated the same security parameters and that the handshake occurred without tampering by an attacker.

ClientHello is the first TLS message transmitted. It includes a session ID, the value of which determines whether the initiated session is a new one or a resumed one. If the client sends an empty session_id it is requesting the server to open a new session; if not, it is attempting to update the security parameters of an older session. A session_id becomes valid at the end of a TLS handshake, after the Finished messages are exchanged. Before that moment, the session ID is neither encrypted nor integrity protected and must not be relied upon for security-related decisions. Through ClientHello, a 28-byte random value is transmitted to calculate fresh keys. In this message, the client also informs the server of the TLS version, the ciphersuites, and compression methods it supports.

If the server can agree on the TLS version and on one of the ciphersuites and compression methods proposed by the client, it will reply by indicating its choice in a *ServerHello* message. If not, it must end the handshake sending an alert. Depending on the session_id value sent by the client, the server will choose an ID for a new session or resume the existing one if it can find a match in its session cache. In the session resumption scenario, in case of success, the peers will jump to exchanging Finished messages; in case of failure, the server will choose a new session_id value to indicate the old session cannot be resumed. Through ServerHello, the server will transmit its 28-byte random value to contribute to the key's freshness.

Server Certificate is an optional message, sent only when server authentication is requested. The certificate generally is, but is not limited to an X.509 type. Certificate chains are supported.

Server Key Exchange Method is an optional message conveying a public key that will be used for premaster secret encryption. It is sent only when server authentication is not supported, implying the Server Certificate message is not transmitted, or when the key contained in the Server Certificate message is not adequate for encryption.

Certificate Request is an optional message sent when client authentication is required. A list of accepted certificate types and certification authorities is included.

The *Server Hello Done* message indicates that the server finished transmitting and that is expecting a response from the client. Upon receiving this information, if a server certificate was received, the client is expected to verify its validity.

The *Client Certificate* is an optional message that should be sent by the client when the server requests client authentication. If no client certificate is available, the server may react with a fatal handshake failure alert.

The *Client Key Exchange* message is always sent by the client to establish a 48-byte premaster secret key. If the chosen ciphersuite uses RSA for public key encryption, the client will pick the premaster secret key and send it encrypted using the server's RSA public key. If Diffie-Hellman was agreed upon, the client will transmit the parameters that will allow agreement on a premaster secret key.

Certificate Verify is only sent when client authentication is necessary. It consists of the client's signature of all previously exchanged handshake structures.

Change Cipher Spec is not part of the handshake protocol, but it's part of the change cipher spec protocol. This message is sent both by the client and server to notify that the newly negotiated parameters and keys will be active from the next message on. The Finished message always follows; it allows verification that the new keys and parameters were correctly set up.

The *Finished* message must be received and validated by the peer before application data can be communicated. Initialization values, encryption, and integrity keys must be available before the Finished message can be computed. They are all obtained by applying a pseudorandom function (PRF) on the premaster secret. The PRF is defined in the TLS standard. The Finished messages are encrypted and integrity protected. The security algorithms are applied to all the handshake data exchanged up to but not including the finished message.

7.1.2 TLS Versus SSL

SSLv3 was the last SSL version created, so that is the one we will compare to TLS v1.0. SSLv1 was fortunately never released, as it contained very serious security flaws, mainly lack of integrity protection as well as of sequence numbers to protect against replay attacks. SSLv2 still contained some security issues and lacked support of desirable functionalities that justified SSL enhancement in v3. The security issues include weak integrity protection and the use of the same keys for message integrity as well as for encryption. Compared to SSLv2, SSLv3 also supports certificate chains and a number of new ciphers, uses SHA-1 instead of MD5 only for integrity protection, and standardized the use of Finished messages to avoid downgrading to a weaker ciphersuite in case of an attacker's interference.

SSLv3 and TLS 1.0 share the same goals and basic protocol structures. In both, a handshake phase has to occur before any application data can be exchanged. The handshake messages are quite similar: main changes involve the ciphersuites supported, key derivation, and message integrity protection calculation.

TLS v1.0 supports a certain number of algorithms that were not included in SSLv3, namely Diffie-Hellman and digital signature standard for authentication, and 3DES for encryption. Adding support for additional ciphersuites is also easier in TLS, as will be shown in Section 7.1.3.

Key derivation is defined differently in the two protocols. TLS defines a standard PRF and applies it. SSLv3 does not define a dedicated function, but rather uses a combination of MD5 and SHA-1 functions. There are no known weaknesses on the key derivation of SSL. Nevertheless, collisions have been found on MD5, which isn't considered secure anymore, and NIST recommends not to use SHA-1 in applications after 2010. These recent discoveries concern TLS security as well as that of SSLv3.

MAC calculation is defined differently in the two protocols. TLS uses the standard HMAC calculation, the security properties of which are well known. SSLv3 applies MD5 and SHA-1 hash functions and a dedicated padding.

When applied, client authentication in SSLv3 is needlessly more complicated than in TLS. Whereas in TLS a hash function is calculated on the handshake messages only, in SSLv3 it is calculated on the handshake messages and the master secret. Explicit use of the master secret is useless, as it is implicitly accounted for in the handshake messages data.

Finished commands are used both in SSLv3 as well as in TLS v1.0. In SSLv3 the handshake integrity is guaranteed by calculating a HMAC on previously exchanged messages, but the input to the HMAC was modified. In standard HMACs, the function is calculated using the key and the data, in this particular order, as inputs. Since in SSL (as well as in TLS), the key is derived halfway through the handshake, for ease of implementation SSL designers proposed applying the HMAC to the handshake concatenated to the key, thus in reverse order. This may weaken the security of HMAC use, so in TLS the standard HMAC function is applied.

Finally, more alert messages were defined and clearly described in TLS v1.0, compared to SSLv3.

7.1.3 TLS Enhancements

The TLS protocol was standardized in 1999; improvements that have been proposed on the original document have been included in separate RFCs [3–5]. At present, a TLS version 1.1 draft has also been published by the IETF as Internet Draft and is under discussion [6].

TLS supports multiple algorithms for authentication, encryption, and integrity protection and was designed so that additional ciphersuites could be easily included. Soon after the release of the TLS specification, Kerberos-based authentication was accepted as an authentication means in TLS handshakes [5]. This was the first attempt to use symmetric cryptography rather than PKI to authenticate the client and the server.

After AES was standardized to replace DES for symmetric key encryption, its use was integrated within TLS [4]. Multiple ciphersuits were defined to use all TLS standard authentication algorithms, namely DH, digital signature standard (DSS), and RSA, with AES for encryption and SHA-1 for integrity protection. In parallel, weaknesses had been discovered in MD5, so none of the proposed ciphersuites used it for hash calculations.

The general extensions to TLS [3] include minor modifications to accommodate hardware constraints on some devices. Most add-ons concern the use of PKI. Clients with memory limitations can send a list of CAs they accept to avoid multiple handshake failures due to their inability to accept the server certificate. They can also use client certificate URLs to prevent occupying memory to store client certificates. Finally, use of the online certificate status protocol (OCSP) response avoids the client consulting long CRLs. The other enhancements involve bandwidth-constrained access networks, making it possible to define maximum fragment length and to use truncated MACs to conserve bandwidth.

TLS v1.1 [6], as described in the Internet draft, will not be substantially different from TLS 1.0 or SSLv3, but it will not be compatible with them either. The main achievements in TLS 1.1 include the following:

- It incorporates the extensions now described in separate RFCs.

- It provides countermeasures against attacks TLS 1.0 was vulnerable to.

- It solves implementation uncertainties that could cause incompatibilities.

- It is purged of the now unnecessary restrictions on algorithm use and key length for exportation.

- It is purged of the patent statement.

PRF calculation and key derivation are equivalent, except that TLS v1.1 explicitly defines initialization vectors (IVs) before encryption starts to avoid a known attack. Also, encryption keys are no longer necessary to fulfill requirements for AES 256-based ciphersuites. TLS v1.1 must be compliant to Internet Assigned Numbers Authority (IANA) definitions on types, including certificate types, ciphersuite types, and alert types.

7.1.4 Shared Key TLS

SSL and TLS decided to rely on a PKI for authentication and for distributing a shared key between a client and a server. The drawback of PKI deployment is its management: a CA must be established for certificate generation, the CA public key must be known for certificate verification, and revocation should be handled to ensure security. These are the reproaches often made to PKI use and that usually justify secret key cryptography instead. Although key distribution and key database management is cumbersome in secret key cryptography, its simplicity justifies its adoption. Also, for equivalent security, public key algorithms are usually more time and resource consuming than secret key cryptography, so when resource-constrained devices are involved or when performance is an issue, it is preferable to use secret key cryptography.

Due to complicated PKI management, as well as the need to use devices that already used Kerberos for authentication, initiatives were soon made to support secret key cryptography for TLS authentication. Kerberos was the first secret key protocol standardized for authentication [5]; other Internet drafts were later published to achieve a similar goal.

Use of Kerberos, though, requires the existence of a trusted third party (TTP) and the ability to connect to it any time authentication between two peers must be achieved. Both peers must trust the same TTP and must previously share a secret key with it, making the protocol less attractive than simple key sharing between the peers.

A few Internet drafts on the use of shared keys for TLS were published in 2004, but they expired and their standardization process was discontinued. At present, only one shared key TLS Internet draft exists, and it is intended for applications in closed environments where secret key distribution can be easily accomplished. It defines three sets of ciphersuites, in which, respectively,

- Only symmetric key algorithms are used for authentication;
- A DH exchange is authenticated with a preshared key;
- Public key cryptography is used for server authentication, while a preshared key is used for client authentication.

The Internet draft defines premaster key computation for each of the three ciphersuite sets. As in the original TLS standard, authentication is achieved by Finished messages verification.

Agreement of a preshared authentication algorithm must occur: the client proposes its use in the ClientHello message and the server agrees on it in the ServerHello message. Because peers must memorize a shared key for each device they want to communicate with, pairwise shared key (PSK) identifiers are needed.

7.2 Other Higher Layer Protocols

In this section, a brief discussion is given of HTTPS, kilobyte SSL (KSSL), and SSH.

HTTPS is basically TLS (RFC 2818) with HTTP running over the TLS protocol instead of HTTP running over TCP. The agent acting as the HTTP client should also act as the TLS client. It should initiate a connection to the server on the appropriate port and then send the TLS ClientHello to begin the

TLS handshake. When the TLS handshake is finished, the client may initiate the first HTTP request. All HTTP data is sent as TLS application data. Normal HTTP behavior, including retained connections, follows. TLS provides a facility for secure connection closure. When a valid closure alert is received, no further data is received on that connection.

KSSL is a client-side–only implementation of SSL v3.0 for handheld and wireless devices. The characteristics of a wireless environment, including weaker CPUs, network latency, low bandwidth, and intermittent connectivity, are taken into account in the application development phase. KSSL is implemented in the J2ME MIDP package (http://developers.sun.com/techtopics/mobility/midp/articles/https). KSSL supports server-side authentication only based on X.509v3 certificates with RSA keys, signed using RSA with MD5 or SHA. It uses RSA_RC4_128_MD5 and RSA_RC4_40_MD5 cipher suites (most commonly used and fast), and only RSA public key operation is used (up to 1,024 bits). SSH is used for remote login from a computer. This protocol provides end-to-end encryption with IDEA or as an option data encryption standard (DES), 3DES, and blowfish, and user and host authentication using RSA [7].

7.3 Known Issues and Possible Solutions

TLS, and SSL before it, have been the main standards for client-to-server communication protection since this need arose. They were designed to provide security to Internet transmissions, where confidentiality was requested for a number of applications and where the need for authentication appeared at first in the mobile commerce and home banking scenarios.

Authentication, and the premaster key establishment that depends upon it, is the protocol's most critical phase. We will only comment on the use of public key authentication since preshared key TLS is still in its development phase.

TLS allows an anonymous communication, server-only authentication, or mutual authentication. TLS anonymous connections provide protection only against passive eavesdropping, not against active attacks. It is needless to say that mutual authentication is recommended to securely connect to a private network (e.g., a corporate network) or for financial transactions. Public key pairs and their relative certificate are usually linked to a device where they are stored, rather than to the device user. This is convenient on the server side, where the goal is to authenticate a corporate, bank, or merchant server, whereas it is totally inadequate on the client side, where we do not want to identify a machine but the person using it. In most commercial TLS applications, a TLS session is opened after server authentication only, and afterward the client is authenticated using other methods, typically login and password. A man-in-the-middle attack could spoof the client to authenticate to it instead of the real EAP server. Once the client credentials are obtained, the man in the middle could forward the victim's authentication credentials to the real server over a protected tunnel. This is possible whenever a protocol is tunneled inside another one without cryptographic binding of the keys in the inner and outer protocols. Solutions are simple: TLS is deployed using mutual authentication or cryptographic binding is enforced.

An entity must verify the peer's certificate before using its public key. The CA that signed the certificate must be among the trusted ones, and the certificate should not have expired. When this is not the case concerning the server's certificate in Web browsing, a pop-up alert should appear[1] to notify the client of the possible risk. Just as with expired documents, expired certificates are not valid anymore and should not be relied upon. Anyone on the Internet can pretend to be a CA and sign certificates; a certificate's worth is directly linked to the CA that delivered it. If the CA is unknown to the peer or if it appears dubious, it should not be trusted and the transaction should be aborted. Users that are inexperienced or in a hurry may be temped to accept any certificate just to get through the handshake phase, greatly endangering their data exchange.

Security requirements are extensively discussed in the TLS RFC [2]. TLS's main defects are its susceptibility to DoS attacks and the nonoptimal construction of its MAC.

Due to the underlying communication technology, usually TCP, TLS is susceptible to DoS attacks. TLS itself cannot do much against DoS attacks; other protocols must be applied if these attacks are to be avoided.

To ensure confidentiality and integrity protection, TLS encrypts transmitted data and calculates a MAC on it. To provide both services simultaneously, security guidelines warn to encrypt and then calculate the MAC on the encrypted text. In TLS, instead, the MAC is computed on the plaintext, and afterward the concatenation of the plaintext and MAC is encrypted. This has been proven secure for certain combinations of encryption functions and MAC functions, but not all of them, so every new ciphersuite must be verified for possible weaknesses.

Finally, TLS PRF and MAC computations are based on both MD5 and SHA-1. Use of both was envisaged so that if one was compromised, the other would still guarantee sufficient security. Unfortunately, attacks have been found on MD5, and NIST recommends not using SHA-1 in applications after 2010.

7.4 Security in WAP

Talking about mobile communications requires at least to some extent an understanding of WTLS. WTLS is used by WAP. WTLS is based on TLS with

---

1. This occurs unless the user explicitly disabled alert messages.

the goal to provide privacy, data integrity, authentication, and reliable communication over the wireless network. The reason for developing WTLS was the resource limitations in wireless devices, memory and processing capabilities, and the wireless medium resource spectrum/bandwidth limitations. Now, WTLS was developed for the WAP 1.x series, whereas the newer version 2.0 was standard TLS [5].

For WAP 1.x, the WAP protocol runs between the mobile terminal and the WAP gateway. At the WAP gateway, the WAP protocol is converted to standard IP stack using SSL. WAP 2.0 uses standard IP stack with TLS at the mobile terminal, thus providing the ability for the TLS tunnel to be created end to end. One of the reasons for the simplification and introduction of IP stack with TLS in mobile terminals was the introduction of iMode in Japan. iMode uses TLS and compressed HTML [8].

In case of WAP 1.x, the gateway thus converts the encrypted traffic in plain text and then reencrypts it for the WTLS connection to the client or the SSL connection to the server. WTLS is very similar to TLS, with additional features that fulfill the requirements for low bandwidth, hazardous nature of wireless medium, peanut processing power, low memory, and cryptographic code export restrictions. To fulfill these requirements, among others, WTLS provides retransmission, duplicate filtering, three levels of authentication including mutual authentication, mechanism for renegotiation of keys, cipher suite including RC5, ECC, and short hash, and small certificate that is simple to parse.

WTLS was found to have several security issues. There are issues such as chosen plaintext data recovery attacks and message forgery attacks. Of course, one of the biggest issues is a compromised WAP gateway.

WTLS can be used with or without PKI/WAP PKI. In case of no PKI, the terminal is hardcoded to connect to the WAP gateway of the mobile network operator (MNO). The MNO connects to the server using SSL from the WAP gateway. With a PKI infrastructure, although rather complex, the WAP gateway can stay opaque and provide connection between the terminal and the server [9].

Certificate provisioning is a major issue in all PKI systems, and an even more sensitive one when mobile platforms are deployed. When public key cryptography is used, certificate verification is mandatory to avoid basic man-in-the-middle attacks. This step implies knowledge of the CA's public key, which should be stored inside the mobile platform before its distribution on the market. This leads to the need to define the CA, or a restricted list of trusted CAs, before the device is in use. This is the most secure approach, but it is also the most restrictive. Another solution may consist of downloading the CA's public key or accessing it from the network (use of uniform resource locators), but this includes a risk factor due to the possibility of relying on an untrustworthy CA.

Unless the mobile platform integrity and its resistance against attacks can be guaranteed, WTLS security is increased when secure storage and crypto-graphic operations are performed on a WAP identity module (WIM). The WIM is a smart card and therefore can offer tamper-resistance characteristics. Its main use is to store the secret keys, perform the calculations requiring the private key, and store the certificates.

WTLS is supported by different OSs, including Symbian.

TLS does not have the same security issues as WTLS; further deploying TLS means the WAP gateway is not required. These reasons, and TLS being off-the-shelf technology, led to choosing it for WAP 2.0.

References

[1]   Freier, A. O., P. Karlton, and P. C. Kocher, "The SSL Protocol version 3.0," Internet Draft, draft-freier-ssl-version3-02.txt, November 18, 1996.

[2]   Dierks, T., and C. Allen, "The TLS Protocol version 1.0," RFC 2246, January 1999.

[3]   Blake-Wilson S., et al., "Transport Layer Security (TLS) Extensions," RFC 3546, June 2003.

[4]   Chown, P., "Advanced Encryption Standard (AES) Ciphersuites for Transport Layer Security (TLS)," RFC 3268, June 2002.

[5]   Medvinsky, A. and M. Hur, "Addition of Kerberos Cipher Suites to Transport Layer Security (TLS)," RFC 2712, October 1999.

[6]   Dierks, T., and E. Rescorla, "The TLS Protocol Version 1.1, Internet Draft," draft-ietf-tls-rfc2246-bis-09.txt, December 2004.

[7]   OpenSSH, http://www.openssh.com.

[8]   Compressed HTML: http://www.w3.org/TR/1998/NOTE-Compact HTML–19980209/.

[9]   WAP specs: http://www.openmobilealliance.org/tech/affiliates/wap/wapindex.html.

8

IP Layer Security

The Internet protocol's (IP's) main goal when it was designed was efficient data transport; no one could have predicted the rapid and wide adoption of the Internet as we know it and the need for support of additional services, including security. The first widely used version of IP was defined in the early 1980s and is known as IPv4. Initially, the TCP combined both TCP and IP functions as we know them today. After evolving through three TCP versions, it was split into TCP and IPv4. This explains the nonexistence of IP versions prior to v4. The need for a successor to IPv4 appeared in the 1990s, so IETF started working on IP Next Generation, also known as IPv6 (also, in this case, v5 does not exist). Most IP applications today still use IPv4, but a number of IPv6 implementations can be found on the Internet. The main need for an upgrade in the protocol is due to IPv4's restricted addressing and the risk of exhaustion. In this chapter, we will focus not on the characteristics of the protocol itself, but rather on its security features.

Security was not a requirement when the Internet was initially designed, but it proved necessary with the development of applications such as mobile commerce and the possibility to remotely access a private network. The design of IP security was also undertaken in the 1990s by IETF. The IPsec standard was designed so that it could be optionally included on an IPv4 network, but it is a mandatory part of the IPv6 structure.

There is no protection on a simple IP packet, making it vulnerable to any attacks concerning:

- *Confidentiality:* Data in clear can be spoofed easily be eavesdroppers on the network.

- *Authentication:* Sender and receiver addresses are not protected and can be modified along the network.
- *Integrity:* Data can be modified during transmission.

IPsec is an IP layer protocol that provides security services, among which are authentication of packet source, data integrity, confidentiality, and antireplay protection. Non-repudiation, access control, and key exchange mechanism are not part of the protocol [1–20].

IPsec includes several protocols that interact to provide overall data security. IP datagram protection is provided using the encapsulating security payload (ESP) or the authentication header (AH). AH provides data integrity, data source verification, and antireplay protection. ESP provides all these services as well as data confidentiality. Key management can be provided in IPsec manually, or thanks to the Internet key exchange (IKE) protocol. IPsec may be implemented in end systems or on security gateways, such as routers and firewalls described in Section 8.5. Two IPsec modes have been defined. Transport mode is used to protect IP payload, but it can only be deployed when data does not have to get through routers or gateways because of encapsulation and decapsulation issues, which will be discussed throughout this chapter. Tunnel mode protects entire IP datagrams and, because of the particular header construction, can be used when packets must transit through security gateways.

A security association defines the processing during a session between two peers, keeping track of security services that must be provided, the packets they must be provided on, and the keys that must be used. For every transmitted packet, this information is indicated in the security parameter index (SPI) in the header. Information on the security association is stored in a security association database (SADB).

Network administrators must specify a traffic security policy to define which security algorithms must be applied depending on the traffic type and on its destination. This information is maintained in the security policy database (SPD). For both inbound and outbound traffic, the policy will define which packets should be discarded, which ones do not need to be security protected, and which ones do. When security must be enforced, the SADB will be consulted.

IP packets can be received out of order, but sequence numbers are used to avoid replay attacks. To limit buffer sizes, a receiving window is defined so that only the packets that are within the window must be accepted and processed.

IPsec is defined in several RFCs [4–7]:

- Security architecture for the Internet protocol (RFC 2401);
- IP security document road map (RFC 2411);

- IP AH (RFC 2402);
- IP ESP (RFC 2406).

8.1 AH and ESP

AH and ESP provide security services within IPsec. Both AH and ESP can work in transport (see Figure 8.1) and tunnel mode (see Figure 8.2). In tunnel mode, a new IP header is added to the packet to designate tunnel sending and receiving ends.

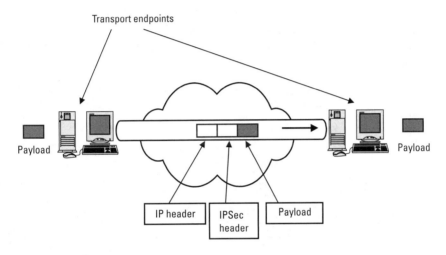

Figure 8.1 IPsec transport mode (headers as in IPv4).

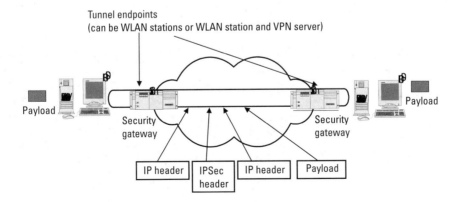

Figure 8.2 IPsec tunnel mode (headers as in IPv4).

AH provides data integrity and data origin authentication as well as replay attack protection. It may provide non-repudiation, but it does not provide confidentiality. AH protects the whole IP packet; fields that would change in transit are not used for calculation of authentication data.

In transport-mode AH, the IP payload and selected header fields are included in the authentication calculation. In tunnel-mode AH, the entire original IP datagram is included in the authentication calculation. The result is placed within a new IP datagram. Selected header fields of this new IP datagram are also included in the authentication calculation. The AH transport and tunnel modes for IPv4 and IPv6 are shown in Figure 8.3 and Figure 8.4, respectively.

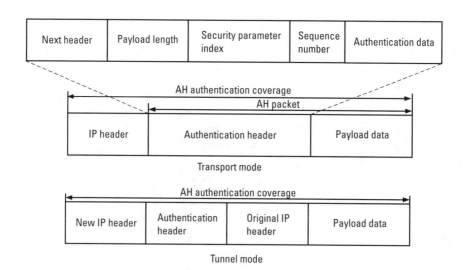

Figure 8.3 IPsec AH transport and tunnel modes packet format for IPv4.

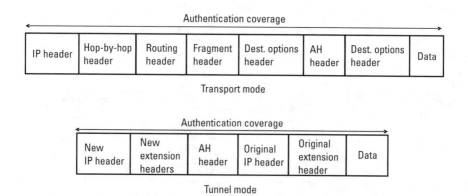

Figure 8.4 IPsec AH transport and tunnel modes packet format for IPv6.

ESP can provide data confidentiality and integrity, data source authentication, and replay attack protection. Protection against traffic analysis is not provided by this mechanism.

ESP can be implemented, providing both encryption and authentication or by providing only one of the services. Unencrypted synchronization data (if required) is carried in the beginning of the payload data field so that the recipient can process the data. This includes the recipient's IP address, the sequence number, and authentication data. Possible encryption algorithms are DES, 3DES, AES, RC5, IDEA, 3-key triple IDEA, CAST, and Blowfish.

In transport-mode ESP, the IP payload (which contains a transport-layer packet) is placed in the encrypted portion of the ESP frame, and that entire ESP frame is placed in the payload of the original IP datagram, the IP headers of which are not encrypted. In tunnel-mode ESP, the entire original IP datagram is placed in the encrypted portion of the ESP, and that entire ESP frame is placed within a new IP datagram having unencrypted IP headers. Encapsulating the protected data is necessary when confidentiality protection is required for the entire original datagram. Figure 8.5 shows the transport and tunnel modes of ESP for IPv4. For IPv6, the extension headers follow the IP header.

8.1.1 Security Policy and Security Association

The security policy (SP) specifies what service is offered to a specific packet based on its selectors: source and destination addresses, protocol, and upper

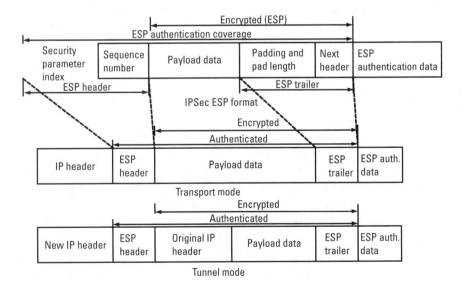

Figure 8.5 IPsec ESP packet format for transport and tunnel modes in IPv4.

layer ports (TCP and UDP). The SP is kept in the SPD. For outbound traffic, the SPD is searched to see if the packet should be protected. For inbound traffic, the SPD is searched to check if the decrypted/authenticated packet was supposed to be protected.

A security association (SA) is the method IPsec uses to track a given communication session. It defines how the communicating systems will use security services, including information about the traffic security protocol, the authentication algorithm, and the encryption algorithm to be used. SAs also contain information on dataflow, lifetime, and lifedata, as well as sequence numbering for antireplay. SAs are negotiated between two IPsec systems. This implies that in tunnel mode SAs are negotiated between two endpoints of the tunnel. The two IPsec systems can also negotiate the level of authorization for a range of addresses, protocols, and ports that will be protected by the SA. An SA is unidirectional; that is, for each pair of communicating systems there are at least two security connections—one from A to B and one from B to A. A given SA can use ESP or AH, not both. If a connection needs both protocols, it needs to establish two SAs for each direction; four for a bidirectional connection. SAs are identified with their SPI. For one peer, an SA is identified with a unique non-ambiguous SPI/<remote peer IP> pair. The SA's main pieces of information are the algorithm to use to protect data (e.g., DES), algorithm-specific attributes (e.g., keys), mode (tunnel/transport), tunnel destination (peer), and proxy identity (selector). A SA is identified by a combination of a SPI, which is a randomly chosen unique number; the destination IP address of the packet; and the traffic security protocol to be used (AH or ESP). Two databases are required for SAs: a SPD that specifies the security services that will be provided for IP packets and a security association database (SAD), in which each entry defines the parameters associated with one SA.

8.2   Key Management

IPsec permits several different key management mechanisms to be used, including manual configuration (pre-established keys). This method also allows separate development and modification of the key management in one hand and AH and ESP protocols on the other hand. The only coupling between the key management protocol and the security protocol is with the SPI.

The key management mechanism is used to negotiate a number of parameters for each SA, including not only the encryption keys but also other information (e.g., the authentication algorithm and its mode) used by the communicating parties. Since manual key distribution is a burden on network administrators, the IKE can be used to establish keys and security associations.

IKE is a family of protocols that is based on the following:

- The Internet security association and key management protocol (ISAKMP), which specifies a framework for key management;
- Parts of Oakley, a key exchange protocol;
- Parts of the secure key exchange mechanism (SKEME), another key exchange protocol;
- Parts of station to station (STS), yet another key exchange protocol.

When the SPD identifies that there is no SA for a communication that must occur, it instructs IKE to create one. IKE will define encryption, integrity, and authentication algorithms and keys. The first operation IKE will perform will be to agree on a suite of algorithms that is acceptable to both peers.

The IKE protocol is divided in two phases: first a shared secret is established, and then it is authenticated. There are several variations of an IKE negotiation, three modes (aggressive mode or main mode for phase one and quick mode for phase two), and three authentication methods (preshared, public key encryption, and public key signature).

During phase 1, the goal of which is to establish a shared secret between peers, IKE negotiates the following:

- How to protect phase 1 (crypto and hash algorithms);
- Hardness of the keys (DH group, since DH is always used to establish the shared secret);
- How to authenticate with the remote peer in step 2 (preshared, public-key encryption, and digital signature);
- Keying material for phase 2.

Overall, IKE ensures that the communication is with the right peer.

During phase 2, the goal of which is to authenticate the peer and create a pair of IPsec SAs, IKE negotiates the following:

- A protection suite (e.g., ESP and AH);
- Algorithms in the protection suite (e.g., DES and SHA);
- Whom we are protecting (proxy identities);
- Optional keying material for negotiated protocols.

A new version of IKE, also known as son of IKE (SOI) or IKEv2 [20], is also being standardized by IETF. IKEv2 can establish IPsec SA in two request/response pairs, as compared to three to four steps in IKE (depending on whether main or aggressive mode was used in phase 1). It is also less complex and thus more secure. On the other hand, the mobile working group is working on enhancements of IKEv2 for multihoming and mobility.

8.3　IP Address Configuration

Let us now look at the issue of IP address configuration in IPsec. Using IPsec as a VPN solution is only beneficial if the user that is physically in an external network can be considered logically as part of the internal network in a secure way. This means that there is a necessity of assigning an internal LAN IP address to the device the user is using while away from the LAN. This is possible roughly based on the following steps:

1. The user device connects to the VPN server and creates an IPsec tunnel.
2. The DHCP address provides the device an internal LAN IP address, which is used as the source address by the device. Here SA is needed with the dynamic host configuration protocol (DHCP) server.
3. The device in IPsec tunnel mode uses the address from the LAN as the internal IP packet source address and the address assigned by the network where it is located as the external IP packet source address.

8.4　Network Address Translation

The major IPv4 problem is the reduced address space, to cope with which *astuces* have to be used to reuse the same address for multiple devices.

Nonroutable private addresses [11] have been defined; these are class A from 10.0.0.0 until 10.255.255.255, class B from 172.16.0.0 until 172.16.255.255, and class C 192.168.0.0 until 192.168.255.255. Private addresses are used for local private networks and are not visible by the public network. When a device in the public network (like the Internet) wants to communicate with a private network address, translation is required to assign it a nonroutable or private address. This is where network address translator (NAT) comes in, so NAT routers (or NATificators) are located at borders of public and private networks. A NAT table is built by mapping private and public addresses. Basically a pool of public IP addresses is shared by an entire private IP subnet.

Address translation can be static or dynamic. In static NAT, a private address is statically linked to a public address. In dynamic NAT, edge devices create bindings on the fly. After the connection is terminated (or a timeout is reached, which is usually short), the binding expires, and the address is returned to the pool for reuse. A variation of dynamic NAT, known as network address port translation (NAPT), may be used to allow many hosts to share a single IP address by multiplexing streams differentiated by TCP/UDP port number.

Translation is not sufficient for connection to a private network address. The payload of the packet must also be considered during the translation process, as checksums and port numbers must be adjusted. NAT must regenerate the IP header checksum, the UDP or TCP header checksum, and the Internet control message protocol (ICMP) header checksum. Also, UDP or TCP port numbers and ICMP message types must be translated.

8.4.1 IPsec and NAT

IPsec, when used together with NAT, faces some issues. In this section, these issues are discussed together with possible solutions.

8.4.1.1 Issues

The first thing that we notice is that there is a change in IP address when using NAT; this means that the integrity in AH will fail. A further change in IP address means recalculation of TCP checksum, which is encrypted in the case of ESP. This is a nonissue for ESP tunnel mode.

Another issue is with IKE and SA setup and endpoint authentication. IKE is based on the IP address as identifier, which must change when NAT is used, but this is always hashed or encrypted. Even in the case where IP addresses are not used in IKE payloads and an IKE negotiation could occur uninterrupted, there is difficulty with retaining the private-to-external address mapping on NAT from the time IKE completes negotiation to the time IPsec uses the key on an application [12]. There are other issues related to SA time-out and IP address time-out.

8.4.1.2 Solutions

There are a few solutions discussed in IETF [13–20]. The most prominent one is the NAT traversal [13], which is discussed here. The main technology behind this solution is UDP encapsulation, wherein the IPsec packet is wrapped inside a UDP/IP header, allowing NAT devices to change IP or port addresses without modifying the inner IPsec packet.

For NAT traversal to work properly, two things must occur. First, the communicating VPN devices must support the same method of UDP encapsulation. Second, all NAT devices along the communication path must be identified.

Usually, NAT assignments last for a short period of time and are then released. For IPsec to work properly, the same NAT assignment needs to remain intact for the duration of the VPN tunnel. NAT traversal accomplishes this by requiring any end point communicating through a NAT device to send a *keep alive* packet, which is a one-byte UDP packet sent periodically to prevent NAT endpoints from being remapped midsession.

All NAT traversal communications occur over UDP port 500. This works well because port 500 is already open for IKE communications in IPsec VPNs, so new holes do not need to be opened in the corporate firewall. This solution does add a bit of overhead to IPsec communications; namely, 200 bytes are added for the phase 1 IKE negotiation and each IPsec packet has about an additional 20 bytes.

8.5 VPN

A VPN connects the components and resources of one network over another network by allowing the user to tunnel through the Internet or another public network, giving the participants the same security and features as those available in private networks [21]. VPNs allow telecommuters, remote employees, or even branch offices to connect in a secure fashion to a corporate server using the routing infrastructure provided by a public internetwork (such as the Internet). Any transport means can be used to transport data, phone lines, dedicated lines, or the Internet, the latter being the most widely deployed choice. A VPN securely transports IP packets across the Internet backbone by establishing tunnel endpoints that share a common encryption and authentication scheme.

The secure connection across the internetwork appears to the user as a private network communication—even though this communication occurs over a public internetwork—hence the name (see Figure 8.6). This solution is financially by far more convenient to implement compared to a dedicated WAN infrastructure.

Some of the common uses of VPN are listed as follows:

- *Remote user access over the Internet:* VPNs provide remote access to corporate resources over the public Internet, while maintaining privacy of information.

- *Connecting networks over the Internet:* The VPN software uses the connection to the local Internet service provider (ISP) to create a virtual private network between the branch office router and the corporate hub router across the Internet.

- *Connecting computers over the Internet:* VPNs allow the department's LAN to be physically connected to the corporate internetwork but separated by a VPN server. The network administrator can ensure that only those users on the corporate internetwork who have appropriate credentials can establish a connection with the VPN server and gain

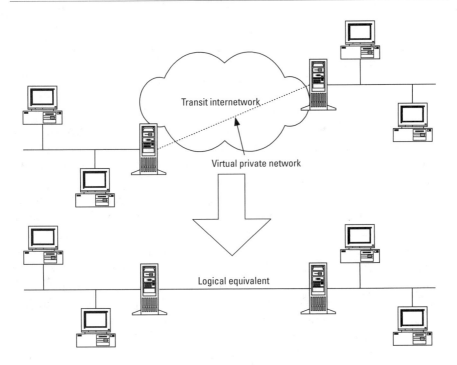

Figure 8.6 VPN and its logical equivalent.

access to the protected resources of the department. All communication across the VPN can be encrypted for data confidentiality.

The requirements for VPN are the following:

- *User authentication:* The solution must verify a user's identity and restrict VPN access to authorized users.
- *Address management:* The solution must assign a client's address on the private net, and must ensure that private addresses are kept private.
- *Data encryption:* Data carried on the public network must be rendered unreadable to unauthorized clients on the network.
- *Key management:* The solution must generate and refresh encryption keys for the client and server.
- *Audit journals:* The VPN gateway may or may not have the ability to create an audit journal of all activities. An audit trail is a chronological record of system activities that is sufficient to enable the reconstruction and examination of the sequence of environments and activities. A

security manager may be able to use an audit trail on the VPN gateway to monitor compliance with a security policy and to gain an understanding of whether only authorized persons have gained access to the network.

VPN fulfills the requirements by making use of tunneling. Tunneling is a method of using an internetwork infrastructure to transfer data from one network over another network. The data to be transferred (or payload) can be the frames (or packets) of another protocol. Instead of sending a frame as it is produced by the originating node, the tunneling protocol encapsulates the frame in an additional header. The additional header provides routing information so that the encapsulated payload can traverse the intermediate internetwork. The encapsulated packets are then routed between tunnel endpoints over the internetwork. The logical path through which the encapsulated packets travel through the internetwork is called a tunnel. Once the encapsulated frames reach their destination on the internetwork, the frame is unencapsulated and forwarded to its final destination. Tunneling includes encapsulation, transmission, and unencapsulation of packets.

IPsec has emerged as the de facto standard for VPNs thanks to its ability to create separate subnets and the extensive review it withstood. All authentication schemes supported by IPsec can be used to connect to a VPN. These include use of login and static or dynamic passwords, tokens, as well as biometric characteristics. For performance matters, VPNs usually use public key protocols for key exchange to establish a shared key and then adopt symmetric algorithms because of shorter encryption time.

Other protocol choices for VPN include the following:

- HTTPS;
- SSH;
- SSL/TLS;
- Layer 2 tunneling protocol (L2TP);
- Point-to-point tunneling protocol (PPTP).

For greater security most enterprises have created their VPN by combining use of firewalls, routers, and one of these protocols.

A router is simply a device that routes data packets through a network; security features, such as encryption or NAT, can be added in higher quality products.

A firewall is a device placed on the network boundaries to discriminate incoming and outgoing traffic depending on access rights. Firewalls accept requests to access network resources only by authorized remote users (i.e., users belonging to the VPN). Firewalls authenticate remote peers, manage key distribution for encryption and integrity purposes, and then allow private data communication. Firewalls must support NAT. They also frequently support encapsulation to hide all packet information that may not be encrypted, including source and destination addresses and headers, and could otherwise leak valuable information.

Performance is low on basic firewalls unless hardware encryption is supported. It is advisable to install personal firewalls on user devices regardless of whether they are from a company IT department or personal. Most of the firewalls, like those from Norton or OutPost, are very simple to install. With firewalls, one can set a different level of security. In general a setting should be such that only the traffic that is needed is allowed to enter and only required data is allowed to leave the system, but this can have some problems. Some applications need to access the network automatically; thus pop-ups might appear often asking permission. Another point is that certain protocols need to make a connection to a user device so as to function, such as a file transfer protocol (FTP) server.

The VPN tunnel can provide a secure connection to users of WLANs. Today corporate users often use VPN. User authentication to the VPN gateway can occur using RADIUS or OTPs. It should be noted that issues like authentication and authorization to enterprise applications are not addressed with VPN. Some VPN devices can use user-specific policies to require authentication before accessing enterprise applications.

VPN organizations and product information can be found at the VPN Consortium [21].

8.6 Mobile IP and IPsec

Today laptops and personal digital assistants (PDAs), or mobile platforms in general that use IP as a transport protocol, expect mobility as one of the functionalities. Mobile Internet protocol (MIP) describes enhancements to IP that allow transparent routing of IP packets to mobile platforms in the Internet. Change of point of attachment to the network is supposed to occur seamlessly, without the user bothering to reconnect and authenticate. MIP allows this to happen today when used over one wireless technology, usually 802.11 or GPRS. In the near future, the possibility to switch from one technology to another should allow running MIP over any technology and to change from one to the other depending on the available and most convenient means.

Traffic is routed over the Internet using source and destination IP address (for IP) and port address (for TCP or eventually UDP). If one of these four numbers change, perhaps because a mobile node changes point of access to the network, the communication is disrupted. In MIP, two IP addresses are used to seamlessly support mobility: each mobile node has a static home address (HA), used to identify the TCP connection, and a dynamic care of address (CoA) that identifies the network to which the mobile node is currently attached. The CoA must be registered with the HA, so that this agent knows where to forward the packets when the device is not attached to its home network. MIPv4 also needs a foreign agent (FA) for neighbor discovery, whereas this functionality is embedded into MIPv6.

This section describes the problems in deploying MIPv4 when working with IPsec-based VPN. There are several ways in which a VPN gateway (GW) and MIP HA can be deployed. Scenarios where IPsec is encapsulated by MIP do not face problems; the issue in such a case is multivendor support. Scenarios where MIP is encapsulated by IPsec have serious problems; this issue will be discussed here.

8.6.1 Scenarios

To start, we will briefly look at different scenarios of deploying MIP and VPN and after that we will discuss in depth the scenario of concern.

Possible ways of deploying MIP and IPsec are listed as follows [22]:

1. *MIPv4 HA(s) inside the intranet behind an IPsec-based VPN gateway:* This requires MIP inside IPsec; this means that traffic between the mobile node (MN) and the VPN server is encrypted. Thus, if a FA is being used it cannot inspect and relay the packet. A CoA might work but it means that the VPN tunnel should be renegotiated every time the MN changes its point of attachment.

2. *VPN GW and MIPv4 HA(s) in parallel at the network border (i.e., VPN and HA are separate):* This scenario can work with MIP in IPsec or IPsec in MIP. MIP in IPsec will have the same problem as in 1. IPsec inside MIP will have no problem, though there will be routing logic modification needed at the VPN gateway or the HA.

3. *Combined VPN GW and MIPv4 HA:* This way, IPsec in MIP can be easily used, but it does not support multivendor interoperability.

4. *MIPv4 HA(s) outside the VPN domain:* Same as 3, except that the HA is separate and placed away from the VPN GW outside the home network.

Combining VPN gateway and MIPv4 HA(s) on the local link, (i.e., using NAT at the firewall and VPN/HA inside the intranet), it can be possible to give the user IPsec connectivity using solutions; now this scenario is similar to 3. In the case of MIP inside IPsec, the problem is the same as in 1.

As scenario 1 is the one supposed to be most practical [22], its issues are further discussed next.

8.6.2 MN Registers with Its MIPv4 HA Using CCoA

Figure 8.7 shows the MIPv4 and the IPsec tunnel endpoints in collocated mode. MN's CoA (most likely obtained through DHCP) is used as both the IPsec and MIP tunnel outer addresses at the MN end.

The MN obtains a CoA at its point of attachment (via DHCP or some other means), and then first sets up an IPsec tunnel to the VPN gateway, after which it can successfully register with its HA through the IPsec tunnel. The problem is that in an end-to-end security model, an IPsec tunnel that terminates at the VPN gateway must protect the IP traffic originating at the MN. If the IPsec tunnel outer address is associated with the CoA, the tunnel SA must be refreshed after each IP subnet handoff, which could have noticeable performance implications on real-time applications. As MIPv6 uses CoA, the issues discussed earlier are also valid for IPsec usage with MIPv6.

8.6.3 MN Registers with its HA Through a FA

Figure 8.8 shows the MIPv4 and the IPsec tunnel endpoints in a hypothetical (but impossible) noncollocated mode. MN's home address and CoA (i.e., a FA address) are used as the IPsec and the MIP tunnel outer addresses, respectively.

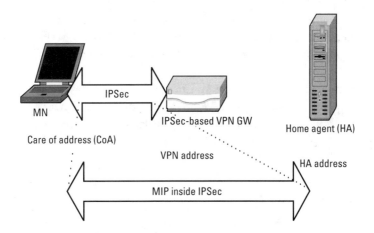

Figure 8.7 MIP with collocated address.

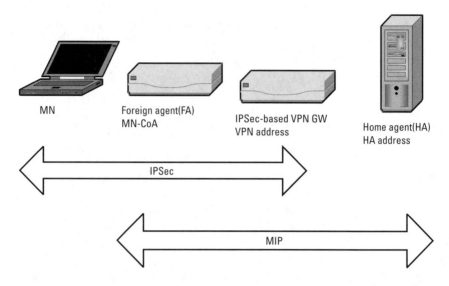

Figure 8.8 MIP with FA.

Please note that the MN does not have a CoA assigned to its physical interface in noncollocated mode.

There are a number of problems with this. Simply put, one could say that the FA needs to see the MIP tunnel outermost, while the VPN-GW needs to see the IPsec tunnel outermost. A more detailed explanation follows.

First, the MN must have an IPsec tunnel established with the VPN-GW in order to reach the HA, which places the IPsec tunnel outside the MIP traffic between MN and HA. The FA (which is likely in a different administrative domain) cannot decrypt MIPv4 packets between the MN and the VPN gateway and will consequently not be able to relay the MIPv4 packets. This is because the MIPv4 headers (which the FA should be able to interpret) will be encrypted and protected by IPsec.

Second, when the MN is communicating with the VPN-GW, an explicit bypass policy for MIP packets is required, so that the MN can hear FA advertisements and send and receive MIP registration packets. Although not a problem in principle, there may be practical problems when VPN and MIP clients from different vendors are used.

8.6.4 Solutions

Reference [23] discusses pros and cons of the solutions available in the open literature. Details will not be given in this section. Solutions discussed in [23] are listed as follows:

1. Dual HA, which says that two HAs should be used, for internal and external, respectively; which leads to three layers of tunnels: external HA, IPsec, and internal HA;

2. The motivation of the next solution, optimized dual HA, is to eliminate use of double MIP encapsulation discussed in 1;

3. Use of MIP signaling to VPN gateway (route optimization);

4. MIP proxy, which aims at introducing a MIP proxy for seamless traversal across VPN;

5. Making VPN GW accept outer IP changes;

6. Use of IPsec instead of Generic Routing Encapsulation (GRE)/IP-IP for MIP tunneling;

7. Host routing and end-to-end security;

8. Explicit signaling to update IPsec endpoint;

9. Use of FA to route ESP.

8.6.5 MIP and NAT Issues

As NAT is often used, MIP's basic assumption fails: MN and FA are uniquely addressable and need global IP addresses.

MIP relies on sending traffic from the home network to the MN or FA through IP-in-IP tunneling. IP nodes, which communicate from behind a NAT are reachable only through the NAT's public address(es). IP-in-IP tunneling does not generally contain enough information to permit unique translation from the common public address(es) to the particular CoA of a mobile node or foreign agent that resides behind the NAT; in particular there are no TCP/UDP port numbers available for a NAT to work with. For this reason, IP-in-IP tunnels cannot in general pass through a NAT, and MIP will not work across a NAT.

MIP's registration request and reply will on the other hand be able to pass through NATs and NAPTs on the MN or FA side, as they are UDP datagrams originated from the inside of the NAT or NAPT. When passing out, they make the NAT set up an address/port mapping through which the registration reply will be able to pass in to the correct recipient. The current MIP protocol does not, however, permit a registration where the mobile node's IP source address is not either the CoA, the HA, or 0.0.0.0.

What is needed is an alternative data tunneling mechanism for MIP that will provide the means needed for NAT devices to do unique mappings so that address translation will work, as well as a registration mechanism that will permit such an alternative tunneling mechanism to be set up when appropriate. A NAT traversal-based solution is discussed in [24].

References

[1]   Black, U., *Internet Security Protocols: Protecting IP Traffic,* Upper Saddle River, NJ: Prentice Hall, 2000.

[2]   Stallings, W., *Cryptography and Network Security: Principles and Practice,* Upper Saddle River, NJ: Prentice Hall, July 1998.

[3]   Frankel, S., *Demystifying IPsec Puzzle,* Norwood, MA: Artech House, 2001.

[4]   Thayer, R., N. Doraswamy, and R. Glenn, "IP Security Document Roadmap," RFC 2411, November 1998.

[5]   Kent, S., and R. Atkinson, "Security Architecture for the Internet Protocol," RFC 2401, November 1998.

[6]   Kent, S., and R. Atkinson, "IP Authentication Header, RFC 2402," November 1998.

[7]   Kent, S., and R. Atkinson, "IP Encapsulating Security Payload (ESP)," RFC 2406, November 1998.

[8]   Maughan, et al., "Internet Security Association and Key Management Protocol (ISAKMP)," RFC 2408, November 1998.

[9]   Harkins, D. C., "The Internet Key Exchange (IKE)," RFC 2409, November 1998.

[10]   Kaufmann, C. (ed.), "Internet Key Exchange (IKEv2) Protocol," Draft, draft-ietf-IPsec-ikev2-13.txt, March 2004.

[11]   Phifer, L., Trouble with NAT, *Cisco IP Journal,* Vol. 3, No. 4, December 2000, pp. 2–13.

[12]   Holdrege, M., and P. Srisuresh, "Protocol Complications with IP Network Address Translation," RFC 3027, January 2001.

[13]   Rosenberg, J., et al., "STUN—Simple Traversal of User Datagram Protocol (UDP) Through Network Address Translators (NATs)," RFC 3489, March 2003.

[14]   Kivinen, T., et al., "Negotiation of NAT Traversal in the IKE," Draft, draft-ietf-IPsec-nat-t-ike-05.txt, January 2003.

[15]   Carpenter, B., and K. Moore, "Connection of IPv6 Domains via IPv4 Clouds," RFC 3056, February 2001.

[16]   Borella, M., et al., "Realm Specific IP: A Framework," RFC 3102, October 2001.

[17]   Borella, M., et al., "Realm Specific IP: Protocol Specification," RFC 3103, October 2001.

[18]   Montenegro, G., and M. Borella, "RSIP Support for End-to-End IPsec," RFC 3104, October 2001.

[19]   Aboba, B., and W. Dixon, "IPsec NAT Compatibility Requirements," RFC 3715, March 2004.

[20]   Eronen, P., and H. Tschofenig, "Extension for EAP Authentication in IKEv2," Draft, draft-eronen-IPsec-ikev2-eap-auth-01.txt.

[21]   VPNC, http://www.vpnc.org.

[22]   Problem "Statement: Mobile IPv4 Traversal of VPN Gateways," Draft, draft-ietf-mobileip- vpn-problem-statement-req-01.

[23]   "Mobile IPv4 Traversal Across IPsec-based VPN Gateways," Draft, draft-ietf-mobileip-vpn-problem-solution-00.

[24]   "Mobile IP NAT/NAPT Traversal Using UDP Tunnelling," Draft, draft-ietf-mobileip-nat-traversal-07.txt.

9

AAA

Human beings have become far more nomadic than ever before. The desire to roam, coupled with the need for connectivity, brought together the development of technology for easy access from anywhere. This need for connectivity for the road warriors led to the development of dialup solutions, a first step toward AAA.

AAA stands for authentication, authorization and accounting standardized by IETF. Today use of AAA has gone well beyond simple dial-up solutions. All network operators, whether mobile or WLANs, use AAA. In this chapter, the purpose and use of AAA will be explained for mobile networks. Before that, though, a brief discussion of AAA solutions will be given [1–18].

9.1 AAA Basics

Before getting into detailed explanation of AAA, in this section the basic explanation of AAA is given. An overview of standardization activities and the simple network architecture is also discussed.

9.1.1 Why AAA?

With the growth of the Internet and the availability and mobility of PCs, the need for remote access arose. This access was given either to offices or to the ISPs to access the Internet. The basic need of such a service was the authentication of the user. The simplest way to authenticate the user was to give secret dialup numbers known only to a closed group of users. This was further enhanced by dial-back and caller ID. The increased demand for Internet access led to the development of network access servers (NASs) that can be distributed at the boundaries of the network. Besides authentication, the user needed to be authorized for a given service in a network. With authorization, the ISP can verify the service the user is allowed to access. Along with this comes accounting, which is based on the extent of service usage. These together form AAA, which should work across different networks and technologies by providing access through an AAA client residing in the NAS and the user profile residing in a database at a centralized AAA server.

Although the basic explanation of authentication and authorization has been touched on in Chapter 1, let us have a brief look at them for completeness purposes. *Authentication* basically means the validation of the identity of the user. It is of utmost importance to validate the identity before the user is allowed access to the network. There are several methods of authenticating a user; methods include username and password pair, shared key, and certificates. Obviously one always runs the risk of authenticating incorrectly. *Authorization*, on the other hand, defines the rights the user has. These rights could be in terms of allowed services (e.g., Web access is allowed but not voice services). In general authorization is done along with authentication. At times authentication means authorization, which is not always the right thing to do. *Accounting* on the other hand leads to billing the user. This information can also be used to plan the network and understand the needs and usage of a given user.

9.1.2   Standardization Activities

The standardization of AAA framework and protocol is the charter of IETF. The current activity of the IETF AAA Working Group (WG) is to develop the DIAMETER protocol, which will have remote authentication dial-in user service (RADIUS) compatibility. Both RADIUS and DIMETER are AAA protocols. Besides IETF, the Internet Research Task Force (IRTF) has an AAA Architecture Research Group.

9.1.3   Network Architecture

In Figure 9.1 the network architecture and network elements associated with AAA are illustrated. There is one AAA server, although there can be several of them. Within the AAA server resides the AAA information about each user. The NAS is located at the edge of the network and contains the AAA client. The basic messages are (1) the user first requests access to the network; (2) the NAS (with AAA client) collects and forwards the user's credentials to the AAA server; (3) the AAA server processes the data and sends an acceptance or rejection to the AAA client; and (4) now the NAS notifies the user device of the success or failure.

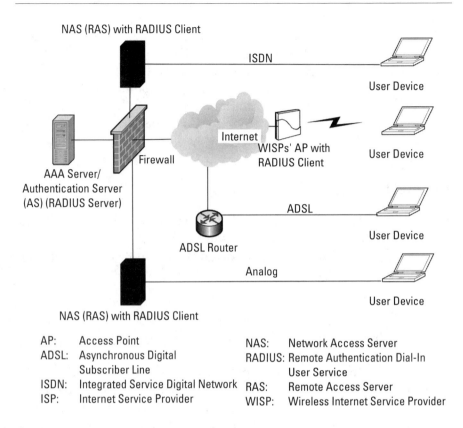

AP:	Access Point	NAS:	Network Access Server
ADSL:	Asynchronous Digital Subscriber Line	RADIUS:	Remote Authentication Dial-In User Service
ISDN:	Integrated Service Digital Network	RAS:	Remote Access Server
ISP:	Internet Service Provider	WISP:	Wireless Internet Service Provider

Figure 9.1 AAA (RADIUS) network architecture.

Together with acceptance or rejection, the AAA server can also send other relevant data, which data could include accounting information. A dialup user will usually interface with a dialup concentrator at the ISP, which then connects to a NAS. The signal of a dialup user makes use of IP over point-to-point protocol (PPP). PPP on the other hand makes use of EAP for user authentication, which itself allows use of other protocols (see Chapter 10). The user device gets the IP address only after authentication.

Usually for authentication, simple protocols like password authentication protocol (PAP) or challenge handshake authentication protocol (CHAP) are used. At the ISP a database like lightweight directory access protocol (LDAP) is used, which contains user account information.

It is possible that the user is not connected directly to its ISP. In such a case, the home network location needs to be found; this is done by using network address identifier (NAI). NAI is basically like an e-mail address: *username@home.network*. This way the intermediary ISP can find the location of the home network. The home network does not need to deploy NAS everywhere; it can rely on other ISPs with which it has contracts/peering agreements.

It is possible that there is a chain of ISPs with such agreements or brokers, known as AAA brokers (AAABs). If an organization is taking a roaming contract from an ISP or broker (e.g., GRIC) then usually the ISP creates one account for the whole organization; this makes management easy for the ISP. Each user of the organization can dial in with the organization's credentials and then create a VPN to the office for enhanced security (see Chapter 8). In Figure 9.2, such a network is given.

9.2   AAA Protocols

There are various protocols defined by IETF. In this section, RADIUS, terminal access controller access control system (TACACS), TACACS+, and DIAMETER are explained. Special attention is given to RADIUS, as it is the most used AAA protocol.

9.2.1   RADIUS

RADIUS is the industry standard protocol for authenticating remote users [1, 2, 4–11]. Today it is widely deployed in remote access servers, routers, and firewalls. RADIUS servers are strategically placed on the network to provide authentication services to all users through a common security protocol. In addition to authenticating and authorizing users, RADIUS enables accounting for the network services. A network configuration of RADIUS is given in Figure 9.1. Key features of RADIUS are

1. *Client/server model:* A NAS operates as a client of RADIUS. The client is responsible for passing user information to designated RADIUS

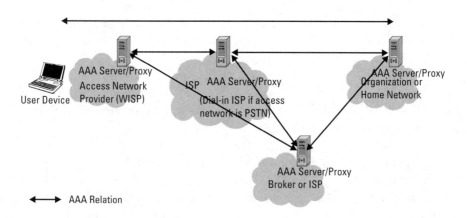

Figure 9.2 AAA for roaming users.

servers and then acting on the response that is returned. RADIUS servers are responsible for receiving the user connection requests, authenticating the user, and then returning all configuration information necessary for the client to deliver the service to the user. A RADIUS server can act as a proxy client to other RADIUS servers or other kinds of authentication servers.

2. *Network security:* Transactions between the client and RADIUS server are authenticated through the use of a shared secret, which is never sent over the network. In addition, any user passwords are sent encrypted between the client and RADIUS server, to eliminate the possibility that someone snooping on an insecure network could determine a user's password.

3. *Flexible authentication mechanisms:* The RADIUS server can support a variety of methods to authenticate a user. When it is provided with the user name and the original password given by the user, it can support PPP with PAP or CHAP, UNIX login, and other authentication mechanisms.

4. *Extensible protocol:* All transactions are comprised of variable length attribute-length-value 3-tuples. New attribute values can be added without disturbing existing implementations of the protocol.

RADIUS starts with the user putting in the username and password in the logon screen; this is passed to the client. The client sends Access-Request to the server, to which the server responds with an Access-Accept or Access-Reject. RADIUS operates over UDP because if a server fails, a secondary server should be queried. TCP is not designed for such purpose; thus, timers and retransmission are kept by RADIUS over UDP.

9.2.1.1 PAP

With PPP, each system may require its peer to authenticate itself using one of two authentication protocols. These are the PAP and the CHAP. When a connection is established, each end can request the other to authenticate itself, regardless of whether it is the caller or the callee. A PPP daemon can ask its peer for authentication by sending yet another LCP configuration request identifying the desired authentication protocol.

PAP works basically the same way as the normal login procedure. (See Figure 9.3.) The client authenticates itself by sending a user name and an (optionally encrypted) password to the server, which the server compares to its secrets database. This technique is vulnerable to eavesdroppers, who may try to obtain the password by listening in on the serial line and to repeated trial and error attacks.

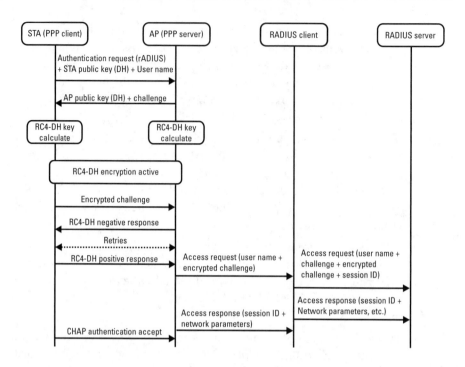

Figure 9.3 RADIUS with PAP over IEEE 802.11 WLAN using RC4 with DH.

9.2.1.2 CHAP

With CHAP, the authenticator (i.e., the server) sends a randomly generated *challenge* string to the client, along with its hostname. (See Figure 9.4.) The client uses the hostname to look up the appropriate secret, combines it with the challenge, and encrypts the string using a one-way hashing function. The result is returned to the server along with the client's hostname. The server now performs the same computation, and alerts the client if it arrives at the same result.

Another feature of CHAP is that it doesn't only require the client to authenticate itself at startup time but sends challenges at regular intervals to make sure the client hasn't been replaced by an intruder, for instance, by just switching phone lines.

The CHAP is used to periodically verify the identity of the peer using a three-way handshake. This is done upon initial link establishment and *may* be repeated anytime after the link has been established.

1. After the link establishment phase is complete, the authenticator sends a challenge message to the peer.

2. The peer responds with a value calculated using a *one-way hash* function.

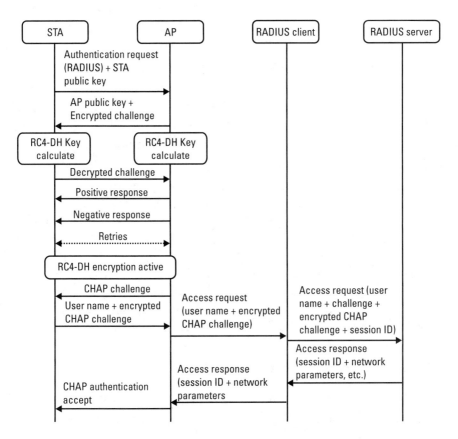

Figure 9.4 RADIUS with CHAP over IEEE 802.11 WLAN using RC4 with DH.

3. The authenticator checks the response against its own calculation of the expected hash value. If the values match, the authentication is acknowledged; otherwise, the connection *should* be terminated.

4. At random intervals, the authenticator sends a new challenge to the peer and repeats steps 1 to 3.

9.2.1.3 Issues

One of the vulnerabilities of RADIUS is the use of a shared secret between the client and the server. If the shared secret is known, there can be lots of threats, such as intruders acting as clients or even servers and collecting user information. Another security issue with RADIUS is that authentication of the access request message is not done, and the use of PAP and CHAP procedures is insecure. Both passive and active attacks are possible against RADIUS. The IETF has proposed solutions for the security issues arising from RADIUS.

In addition, RADIUS is limited due to its command and attribute address space structure, leading to restrictions on introducing new services. Further, RADIUS assumes that there are no unsolicited messages from the server to client, which restricts its flexibility.

9.2.2 Diameter

Diameter can be considered as a next generation RADIUS protocol [15]. It has been developed to address RADIUS flaws in interdomain roaming support and to provide a much more scalable architecture. Its framework consists of a base protocol and a set of protocol extensions (e.g., end-to-end security, PPP, MIP, and accounting). The base protocol provides all the basic functionalities that must be provided to all the services supported in diameter, while application-specific functionalities are provided through extension mechanisms. The most important difference between diameter and RADIUS is that diameter is based on a peer-to-peer architecture instead of client/server model. This easily allows service providers to cross-authenticate their users and to support mobility between many different domains. MIP with diameter is discussed in Section 7.4.

9.2.3 TACACS/TACACS+

TACACS was the first protocol designed to allow remote access with user name and password [12]. The protocol was reengineered by vendors, and the latest version was known as TACACS+. The services provided by TACACS and RADIUS are similar except that TACACS+ works over TCP; it encrypts the entire payload while RADIUS encrypts the user password only, and, unlike RADIUS, it separates authentication and authorization.

9.3 MIP and AAA

Information on how MIP should work with an AAA server is given in [13, 14]. The network elements involved in a handover using MIP and AAA are given in Figure 9.5. Note that this kind of handover will be necessary when the user moves from one administrative domain to another.

There are three AAA elements: the AAA home (AAAH), the AAA foreign (AAAF), and the AAAB. It is possible that AAAH and AAAF have some prior relation (e.g., roaming contract between two ISPs); in that case AAAB is not required. When there is no prior trust relationship between AAAH and AAAF, then it is possible for them to have a trust relationship with an AAAB. In that case, AAAB can create a trust relation between AAAF and AAAH. The basic point about trust is that there is a SA between two network elements.

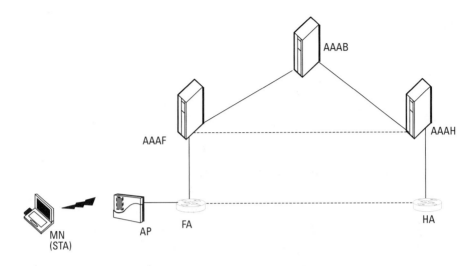

Figure 9.5 MIP and AAA interaction.

In Figure 9.5, there is SA1 between AAAF and AAAB, SA2 between AAAH and AAAB, and SA3 between AAAH and AAAF. SA3 can be created through AAAB as explained earlier, or it can previously exist. There is also SA4 between FA and AAAF, SA5 between HA and AAAH, and SA6 between MN and HA. Now when the MN requests registration at the FA, the FA will send the request to the AAAF that will start the AAA protocol. Of course, this means that the user will be involved, and thus there is no possibility of service continuity.

The easiest way of course would be that the MN on authentication request sends its NAI. Using the NAI, the AAAF can find the AAAH. Before any negotiations can take place between AAAF and AAAH, the SA3 should be created. The AAAH verifies the authentication response and then passes the registration request to the HA. The HA then sends the registration response to the AAAH, which in turn proxies it together with an authentication response to the AAAF. The AAAF then authenticates the user and sends the registration response through the FA. Now the communication between HA and FA can take place. During this exchange a SA (SA7) is also created between FA and HA.

9.4 Wireless Internet Service Provider Roaming

Roaming between public WLAN networks is a goal to achieve a bigger footprint for small wireless ISPs (WISPs). The Wi-Fi Alliance has developed a recommendation for roaming between WISPs known as WISP roaming (WISPr) [16]. This recommendation proposes the use of universal access method (UAM). In this section, basic methods for roaming and the WISPr recommendation are briefly explained.

9.4.1   Inter-WISP Roaming Methods

To roam between WISPs, there has to be some business relationship. There are three basic ways this can be done:

1. In an inter-WISP relationship, each WISP has a roaming contract with every other WISP. Thus, a user of one WISP can roam to another and still receive one bill. This of course means too many separate contracts that can lead to overloading of the network as the number of roaming users increases.

2. In the roaming consortium approach, a consortium is built of which different WISPs become members. Such a consortium can set the contract relationship between different WISPs and can also act as a clearinghouse.

3. The broker method is perhaps the most flexible approach. Here different WISPs have a contract with a broker that allows roaming from one WISP to another depending on such factors as the service level agreement (SLA). The broker can be in the position to authorize a user or if needed pass the user credentials to the appropriate WISP's AAA. This is the most flexible method for roaming.

9.4.2   UAM and WISPr

UAM is the browser-based user authentication and authorization method used widely in many public hotspots. With this method, any IP-based device with a Web browser that supports SSL can login and be authenticated to the hotspot network. The network basically consists of a STA, which communicates through an AP to a DHCP server (if an IP address is needed), to a public access controller, a Web server, and an AAA server.

After the STA is associated, it is given an IP address. Now the user starts the Web browser that leads to a HTTP request. The HTTP request is captured by the Public Access Control (PAC) and sent to the Web browser that displays a logon page to the user. This also starts a SSL connection. The user then types the username and password, which are passed to the AAA server. On authentication, the AC is informed. IEEE 802.11i using IEEE 802.1X and EAP methods can also be used to give a higher level of security.

So to provide roaming with UAM, WISPr recommends a roaming intermediary. The roaming intermediary is like the AAAB.

9.5 AAA in Mobile Systems

AAA is now also being used in mobile communications systems. During standardization process, 3GPP2 and 3GPP communicated with IETF concerning their requirements. Diameter has taken these requirements in account.

9.5.1 3GPP2

In 3GPP2, AAA is used in packet switched mode. Basically a packet data serving node (PDSN) is defined that sends the AAA signal to the AAA server, and on authentication allows users to access the IP services.

9.5.2 3GPP

Similar to 3GPP2, the use of AAA arises only when packet switched services are of concern; for example, the session initiation protocol (SIP) makes use of AAA services with 3GPP. A big role of AAA in 3GPP is visible for WLAN interworking.

An architecture for integration of WLAN and 3GPP access networks belonging to different stakeholders is given in this section with supports to the following solutions [17, 18]:

1. SIM-based authentication for mobile postpaid and prepaid and roaming users;
2. For prepaid users, real-time charging and billing;
3. Short message service (SMS) using OTP for postpaid and prepaid users;
4. Subscription-based billing using username/password credentials;
5. Support for Internet roaming users.

Secure mobility across different stakeholders is achieved using IPSec with MIP.

The architecture as shown in Figure 9.6 is also compliant with the six levels of interworking that have been defined by 3GPP, spanning from the simple common billing (scenario 1) to the seamless service continuity when moving from the 3GPP access network to the WLAN access network, and vice versa, as shown in Table 9.1.

The intention of scenario 3 is to provide access to all 3GPP packet switched–based services: the ones that are available now over GPRS, and the ones that in the future will be provided for packet switched access, namely IMS-based services. The architecture considered here is based on end-to-end VPN tunneling from WLAN user equipment (UE) to the packet data gateway (PDG), which can

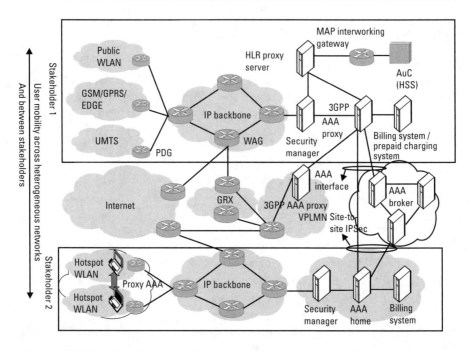

Figure 9.6 3G-WLAN deployment architecture.

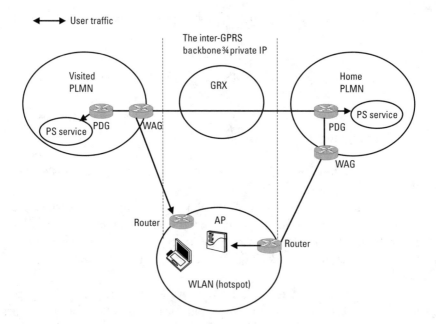

Figure 9.7 End-to-end tunneling architecture.

Table 9.1

Levels of Interworking Between a WLAN and a 3GPP Network

Scenario	1 Common Billing and Customer Care	2 3GPP System-Based Access Control and Charging	3 Access to 3GPP System Packet Switched-Based Services	4 Service Continuity	5 Seamless Services	6 Access to 3GPP System CS-Based Services
Services and operational capabilities	X	X	X	X	X	X
Common billing	X	X	X	X	X	X
Common customer care		X	X	X	X	X
3GPP system-based access control		X	X	X	X	X
3GPP system-based access charging			X	X	X	X
Access to 3GPP system packet switched-based services from WLAN				X	X	X
Service continuity					X	X
Seamless service continuity						X
Access to 3GPP system with seamless mobility						X

be located in the home or visited network depending on where the service is provisioned. The WLAN access gateway that enforces the routing of the user traffic from the WLAN access network to the PDG (in the case of roaming through the interoperator interface/network), is located in the home network in case of home-service access, and in the visited network in case of visited-service access, as shown in Figure 9.7.

Although scenario 3 is not supposed to deal with mobility issues, it is understood that the chosen tunneling solution needs to be future-proof, in the sense that it must be possible to migrate to mobility scenarios 4 and 5 without changing the standardized architecture of scenario 3, possibly by using off-the-shelf MIP solutions. The support of seamless mobility may also depend on the terminal capacity of being able to connect to WLAN and GPRS/UMTS simultaneously.

Note that recent developments in 3GPP can lead to changes in the scenarios of Table 9.1.

References

[1]   Smith, R.E., *Authentication: From Passwords to Public Keys*, Boston: Addison Wesley, 2002.

[2]   Metz, C., "AAA Protocols: Authentication, Authorization, and Accounting for the Internet," *IEEE Internet Computing*, Nov.–Dec. 1999, pp. 75–79.

[3]   IEEE 802.1X, "IEEE Standard for Local and Metropolitan Area Networks—Port-Based Network Access Control," July 2001.

[4]   "PPP PAP and CHAP," RFC 1334, October 1992.

[5]   "CHAP," RFC 1994, August 1996.

[6]   "Remote Authentication Dial-In User Service (RADIUS)," RFC 2865, June 2000.

[7]   "RADIUS Accounting," RFC 2866, June 2000.

[8]   "RADIUS Accounting for Tunneling," RFC 2867, June 2000.

[9]   "RADIUS Authentication for Tunneling RJFC2869 RADIUS Extensions," RFC 2868, June 2000.

[10]   "RADIUS over IP6," RFC 3162, August 2001.

[11]   "Microsoft Vendor-Specific R ADIUS Attributes," RFC 2548, March 1999.

[12]   "TACACS," RFC 1492, July 1993.

[13]   Glass, S., et al., "Mobile IP Authentication, Authorization, and Accounting Requirements," RFC 2977, October 2000.

[14]   Dommety, G., et al., "AAA Requirements from Mobile IP," http://www.ietf.org/proceedings/99jul/slides/mobileip-aaa-99jul.pdf.

[15]   Calhoun, P., et al., "Diameter Base Protocol," RFC 3588, September 2003.

[16]   Anton, B., B. Bullock, and J. Short, "Best Current Practices for Wireless Internet Service Provider (WISP) Roaming," v1.0, Wi-Fi Alliance, February 2003.

[17]   Prasad, N. R., "Adaptive Security for Heterogeneous Networks," Ph.D. Thesis, University of Rome, Tor Vergata, Rome, Italy, 2004.

[18]   3G Security, "Wireless Local Area Network (WLAN) Interworking Security," 3GPP TS 33.234 V1.0.0 (2003-12), Release 6.

10

IEEE 802.1X and EAP

IEEE has defined a generic means of authenticating and authorizing devices attached to a local area network (LAN) or metropolitan area network (MAN) port with point-to-point connection characteristics, regardless of the specific communication technology applied. The goal of port-based network access control is to prevent access to a port prior to authentication and authorization, and in cases in which these processes fail. Authentication and authorization are necessary because of the environments in which LANs are deployed, where unauthorized devices could be connected to the LAN or where unauthorized users could attempt to login on a machine connected to the LAN. A typical scenario is a university campus, hotel, or public building LAN or MAN, where access to resources and services must be granted only to authorized users and devices.

EAP is the wrapper that 802.1X uses to exchange authentication messages. EAP defines packet format and its transmission but does not identify any specific protocol to use. Multiple EAP methods, compatible with EAP general format, have been designed to meet specific needs. The main ones will be described in this chapter.

The 802.1X standard, or port-based network access control, was initially defined for local and metropolitan area wired networks such as Ethernet and token ring. The specification was then updated in 2001 to extend its use to wireless communication systems such as 802.11 and 802.16. Authentication and authorization to a WLAN or WMAN is crucial, as an unauthorized user doesn't need physical access to the network to attempt accessing its resources.

Herein we discuss 802.1X and EAP security because many wired as well as wireless technologies today do not rely on a dedicated authentication protocol but rather refer to the 802.1X standard. All aspects of authentication have already been tackled in 802.1X, and, by relying on a long-existing standard, developers have greater confidence that a new attack will not break their systems. Manufacturers can reuse parts of their existing designs to provide authentication in different technologies. Finally, because of the growing interaction between wireless technologies, the use of a single broad authentication protocol should ease handover (e.g., between 802.11 and 802.16 technologies).

10.1 IEEE 802.1X

In the context of the 802.1X standard, a port is defined as a single point of attachment to the network. Practical examples include the point of attachment of a server or router to a LAN, as well as the association between a station and an access point in 802.11 networks.

The standard provides a means to perform authentication and authorization. Authentication is the process that allows the determination of whether a user or device has the right to connect to a network. Authorization defines whether a user or device belongs to a network. Authentication should therefore be requested before authorization can be granted.

IEEE 802.1X or port-based network access control was designed to provide higher layer authentication mechanisms to layer 2 [1]. Basically, IEEE 802.1X has three entities (see Figure 10.1):

- *Supplicant:* The device desiring to join the network, in our case, the IEEE 802.11 station.

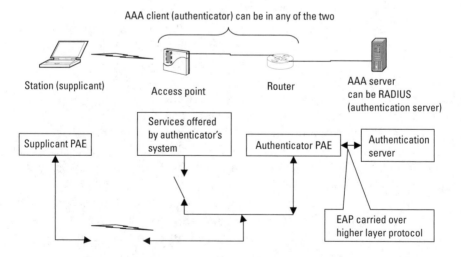

Figure 10.1 Roles of supplicant, authenticator, and authentication server in IEEE 802.1X.

- *Authenticator:* The device that controls the access; in a WLAN network it can be the IEEE 802.11 AP or the access router (AR).

- *Authentication server:* This makes the authentication decision (e.g., the RADIUS server).

The specification defines the principles of operation of the access control mechanisms, the supported levels of access control, as well as the communication protocol between the supplicant and the authenticator and between the authenticator and the authentication server. Message format, timing, retransmission, message transmission state machine, as well as the management protocol are described in detail.

The point where the supplicant connects to a network via the authenticator is called the port or port access entity, thus the designation *port based.* Although the point of attachment to the LAN is single, there are two logical ports controlled by the authenticator: a controlled port and an uncontrolled one. When a supplicant first connects, it goes through the controlled authenticator port to the authentication server. At this point, the authenticator only accepts authentication frames or requests to access services that are not subject to access restrictions. Once the authentication is successful for the supplicant, the service's port is made available, and any data frame can get through the uncontrolled port. Now the supplicant can access the services through the authenticator.

The protocol that 802.1X uses for communication between the supplicant and the authenticator is EAP. Since transmission occurs over a LAN, the protocol used is called EAP over LAN (EAPOL) [1].

The authenticator and the authentication server may or may not be collocated. In case the authentication server is at a remote location, RADIUS may be used to transport EAP messages between the authenticator and the authentication server.

The EAPOL messages of concern for IEEE 802.11 are listed as follows (the message sequence is shown in Figure 10.2):

- *EAPOL-Start:* Determines whether there is an authenticator. Used by sending this message to a special group multicast to MAC address reserved for 802.1X authenticator. Response is an EAPOL-Identity Request in EAPOL-Packet.

- *EAPOL-Key:* Authenticator sends encryption keys to the supplicant.

- *EAPOL-Packet:* A container for transferring EAP messages on LAN.

- *EAPOL-Logoff:* Disconnection message.

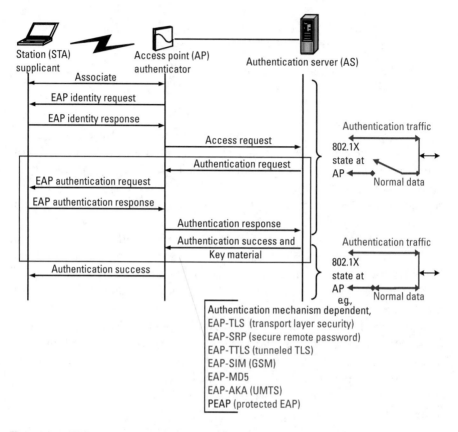

Figure 10.2 IEEE 802.1X message sequence.

10.2 EAP

EAP was designed to solve a major problem, the assignment of an IP address after authentication in an IP network [2]. IPSec and SSL run on an IP layer with knowledge of the IP address. Today EAP has become an important part for WLAN. EAP can be used over layer 2, over IP, or over any other higher layer; it was designed as an extension of PPP.

EAP was not defined with the intent to provide authentication; it is only a wrapper that gives flexibility in usage of any kind of authentication protocol. Thus, an AP does not need to know all the kinds of authentication protocols. Whenever a communication technology relies on 802.1X, and therefore EAP, for authentication, the goal of the protocol usually is to allow authentication and to distribute a shared key, generally referred to as a master key, between the supplicant and the authentication server.

EAP is an IEFT RFC that was standardized in 1998 as RFC 2284 [2]. RFC 23748 has been published and supersedes RFC 2284. An EAP packet is composed of four fields: code, identifier, length, and data. Originally, only four possible packet types were defined: request, response, success, and failure. A communication using EAP was characterized by the bidirectional transmission of request and response frames, finally ending with a success or failure notice. RFC 2284bis includes additional features and defines more EAP types, defines retransmission in details, and thoroughly describes security considerations. RFC 2284bis allows for support of sequences of authentication methods, as occurs for example with protected EAP (PEAP) [3]. A pass-through behavior mode was added to allow the authenticator to transparently forward messages to the authentication server. A peer-to-peer operation mode was also included to allow a device to act both as supplicant and authenticator to support independent and simultaneous mutual authentication.

10.2.1 EAP Security

The security level offered greatly changes according to the EAP method deployed. We will herein list general EAP security considerations and detail the pros and cons of specific EAP methods in dedicated sections.

An identity exchange is optional within the EAP conversation. It is also possible for the identity in the identity response to be different from the identity authenticated by the EAP method. This may be the sign of an occurring attack, but it could also be intentional to conceal the peer's real identity. An EAP method should nevertheless use the authenticated identity when making access control decisions.

To solve the identity privacy issue, it has recently been proposed to tunnel one EAP method inside another one, as occurs in PEAP. This allows the generation of an encrypted tunnel prior to the transmission of identity information. It has been shown that man-in-the-middle attacks are possible within tunneled EAP methods when cryptographic binding between the two methods' keys is not implemented [3]. EAP does not permit untunneled sequences of authentication methods to avoid man-in-the-middle and replay attacks. Tunneling EAP within another protocol enables an attack by a rogue EAP authenticator to tunnel EAP to a legitimate server and should consequently be avoided.

Basic EAP supports per-packet data origin authentication, integrity and replay protection, but false EAP packets could still be injected or replayed. Also, EAP headers are not protected. Some specific EAP methods support integrity and replay protections and should be preferred.

It is not advised to use EAP methods that use algorithms vulnerable to specific attacks. EAP-MD5, for example, is vulnerable to dictionary attacks and should not be deployed for this reason.

EAP does not mandate mutual authentication: no authentication or one-way authentication are accepted but should not be used because of the risk of connection to a rogue device.

It is possible to use an EAP method for a client and a server to derive a shared key. In this case, it is mandatory for the devices to mutually authenticate each other before deriving the key.

If a peer accepts multiple EAP methods, negotiation attacks in which the attacker negotiates the least secure method are possible. To avoid negotiation attacks it is suggested for a peer to propose a single EAP method. If different EAP methods can be used under different circumstances, use a different identity for each accepted method.

Within EAP, success and failure packets are neither acknowledged nor integrity protected. Although results themselves are not protected, a method providing integrity protection and replay protection is less vulnerable to attacks, but DoS attacks are still possible in most cases.

10.2.2 EAP Methods

Due to a lack of space, a detailed explanation of the discussed EAP protocols is not possible; the authors hope that the message sequence charts help the readers to understand the protocols. The protocol stack of EAP is shown in Figure 10.3.

10.2.2.1 EAP-TLS

The EAP-TLS [5] procedure is basically the SSL/TLS procedure shown in Figure 10.4 wrapped in an EAPOL packet. After the optional exchange of identity EAP request and response messages, the authentication server requests to perform EAP-TLS as authentication method. The supplicant shall respond by sending an EAP response containing the TLS ClientHelloMessage. From there on, authenticator and supplicant will continue the EAP exchange, in which the EAP frames format the TLS sequence. After the last TLS message is sent, namely the authenticator's Finished message, the authenticator will send an EAP success or failure message to state whether or not the TLS authentication was successful.

EAP-TLS features and its security are equivalent to that of TLS. For example, just as in TLS, no authentication, one-way, or mutual authentication may be implemented. Also, any among TLS's ciphersuits may be applied to perform EAP-TLS.

To obtain adequate security, mutual authentication should be mandatory, as well as certificate verification at both the client and server side. There should be no security breaches due to the transmission of EAP messages through the authenticator to the authentication server when they are not collocated. If RADIUS is used between them, their communication may be encrypted by the AP-RADIUS key.

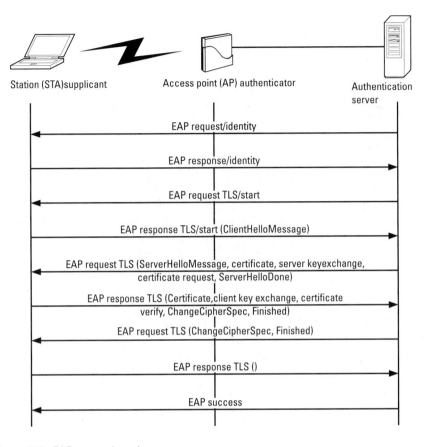

Figure 10.3 EAP protocol stack.

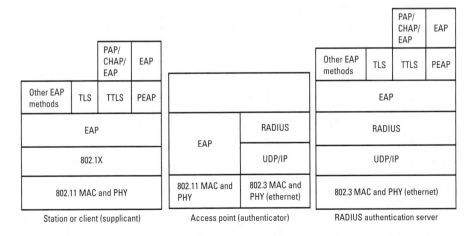

Figure 10.4 EAP-TLS.

The use of a client certificate that is not understood by the end user, lack of user identity protection, and unprotected EAP-success/fail messages are drawbacks or weaknesses of EAP-TLS. This has led to development of EAP tunneled TLS (TTLS) and PEAP. These two are explained in the next section.

EAP-TLS provides mutual authentication and key generation but lacks a solution for supplicant identity protection.

10.2.2.2 PEAP

The lack of privacy in EAP-TLS caused by sending the identity of the user in the open is the issue that PEAP tries to solve. Protected EAP (PEAP) allows tunneling of multiple EAP methods, one wrapped inside the other.

PEAP can be logically divided in two phases, each one corresponding to the execution of an EAP method. The outer EAP method executed is equivalent to EAP-TLS, in which only the server is authenticated. After server authentication, its goal is for the peers to agree on an encryption key to initiate a secure communication. Once privacy is achieved, the inner EAP method is executed. The goal of this second phase is client authentication. In the following paragraphs, an example of the two phases is discussed. The message sequence chart (MSC) is shown in Figure 10.5 [4].

In phase 1 the normal TLS is used, except that the user does not send a username; instead, it sends an arbitrary name. Usually this name will contain information to identify the backend authentication server; thus, a normal NAI is used (e.g., anonymous@companyname.com) [6]. The server sends its certificate in this phase, but the client does not have to do so or else the privacy issue remains, and the procedure simply becomes the same as EAP-TLS.

After phase 1 the protocol automatically starts phase 2. In phase 2, the protocol restarts with the user identity part. In phase 2, any EAP method can be used. Note that the user identity in phase 2 is the real one and is not compared with that in phase 1. In this phase the communication is encrypted by using the keys created during phase 1, so the client's identity can be concealed.

The major security concern with PEAP is possibility of man-in-the-middle attacks. Since there is no client authentication in phase 1, the server sets up an encrypted channel with a device whose identity is unknown until phase 2 is performed. Rogue attacks can be performed in this scenario. Man-in-the-middle attacks can be avoided by employing key binding between phase 1 and phase 2.

PEAP provides mutual authentication, supplicant identity protection, and key generation.

10.2.2.3 EAP-TTLS

Development of EAP-TTLS [7] started with the thought of leaving the legacy systems untouched and still providing the required level of security. The solution was to introduce a TTLS server that lies at the hotspot network or at an ISP

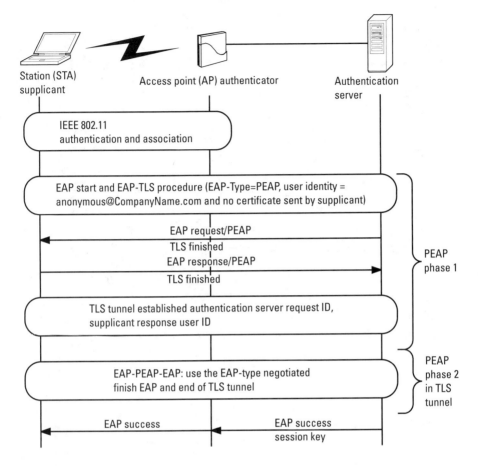

Figure 10.5 Example of PEAP MSC.

while the AAA server is in the home network. Besides that, backward compatibility is achieved by using attribute-value pairs (AVPs), which are compatible with both RADIUS and diameter. EAP-TTLS also consists of two phases which will be explained later. The MSC is given in Figure 10.6.

In the first phase, similar to PEAP, an EAP-TLS secure channel is created. The client side certificate and thus client authentication is optional.

In phase 2 a secure tunnel is established between the client and the TTLS server. Separate protection must be provided between the TTLS server and the home AAA server. The client sends the AVPs to the TTLS server, which checks whether the sequence of AVPs includes authentication information and forwards the information to the home AAA server.

TTLS provides mutual authentication, supplicant identity protection, key generation, and data cipher suite negotiation.

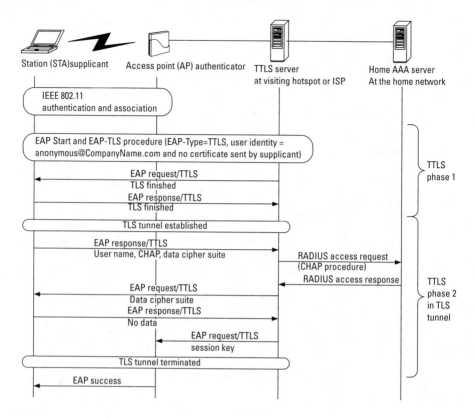

Figure 10.6 Example of EAP-TTLS MSC.

10.2.2.4 EAP-FAST

EAP-flexible authentication via secure tunneling (FAST) [8] is a relatively new proposal in IETF. The basic idea of this protocol is to avoid the usage of certificates. EAP-FAST tunnel establishment relies on a protected access credential (PAC) that can be managed dynamically by EAP-FAST through the AAA server. This method also has two phases with an optional phase 0.

The optional phase 0 is used infrequently. In this phase, per-user credential is securely generated between the user and the network. This credential, known as PAC, is used in phase 1.

Phase 1 establishes a tunnel between the station and the AAA server. PAC is used for authentication purposes.

The tunnel created in phase 1 is used in phase 2 to securely perform client authentication. The user sends the username and password in this phase.

EAP-FAST is said to provide protection against man-in-the-middle attacks, weak IV attacks, replay attacks, and dictionary attacks.

10.2.2.5 EAP-SIM

A number of mobile operators are already providing WLAN access. The method for them to do so is to reuse their current infrastructure. The current infrastructure of operators makes use of SIM-based authentication [9]. The MSC is given in Figure 10.7.

EAP-SIM is based on GSM authentication, so to run this EAP method a user must own a SIM in which *Ki*, the key shared between a GSM user and a mobile operator, is stored. The EAP-SIM protocol may be developed entirely inside the SIM, or else the SIM could be accessed only to retrieve the results of the GSM authentication. When access to a WLAN or WMAN is obtained through EAP-SIM, no mobile voice communications are involved and so the key Kc, initially defined to encrypt voice in GSM systems, can be used to calculate message authentication codes for network and user authentication, as well as an 802.11 encryption specific key.

To execute EAP-SIM, multiple GSM authentications must be performed. After the calculation of *n* GSM authentication triplets (RAND, SRES, Kc), an EAP-SIM master key is derived from *n Kc* values. The master key will be used in

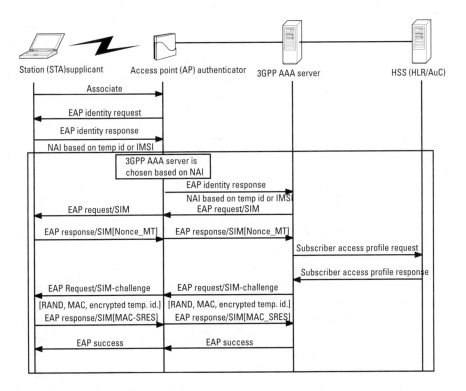

Figure 10.7 EAP-SIM MSC.

the generation of message authentication code keys for peer authentication, for encryption and integrity keys for the method's protection, as well as to generate a shared key between the supplicant and the authenticator for technology-specific security issues.

There are multiple security issues related to the EAP-SIM method. First of all, GSM security is defined only over the air between a mobile user and his receiving base station. Consequentially, EAP-SIM cannot provide security inside an operator's network or between operators, an area where data is not necessarily protected and where attacks cannot be blocked.

Even in the communication range where security should be provided, authentication replay attacks can easily be performed. Since the EAP master key depends on K_c values as unique keys, if K_c values are known a priori, authentication can be replayed. The user does not have means to verify that the server is using fresh K_c values, so server authentication is doubtful because the server only chooses GSM triplets (and consequently K_c values). User authentication isn't very secure as well because an eavesdropper could have memorized a victim's K_c values, although this scenario is less likely because the server is supposed to choose random triplets at every authentication.

It is also worth mentioning that the GSM Kc key is only 64 bits long, so no matter how long EAP-SIM keys are, their ultimate strength can never surpass 64 bits.

EAP SIM provides mutual authentication, although authentication weaknesses exist and have been described; it allows the generation of a shared key between the supplicant and the authenticator. Although EAP SIM uses temporary identities and pseudonyms, there are certain cases in which identity protection cannot be deployed.

10.2.2.6 EAP-AKA

The AKA protocol is used by the third generation (3G) standard developed by the 3GPP. It is based on symmetric keys and runs on universal mobile telecommunications systems (UMTS) SIM or USIM. EAP-AKA [10] was developed so that WLAN users could be authenticated by a 3GPP network. The MSC is given in Figure 10.8.

EAP-AKA formats 3GPP AKA security into EAP format. Similarly to EAP SIM, a 3GPP AKA protocol is performed to generate an EAP master key shared by the supplicant and the authenticator. Based on this key, mutual authentication is performed, EAP encryption and identity protection keys are calculated, and a key shared between the supplicant and the authenticator is calculated for technology-specific security issues.

EAP-AKA solves most of the security issues that affect EAP-SIM. As 3GPP security is defined within and between operator's networks, there are no areas where data is transmitted in clear. 3GPP defines mutual authentication

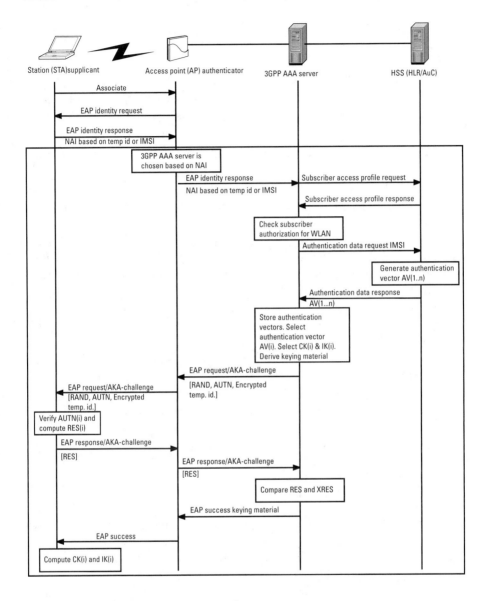

Figure 10.8 EAP-AKA message sequence chart.

between the network and the user, allowing it to be provided also in the scope of EAP-AKA; the use of counters counteracts replay attacks. Because 3GPP keys are 128 bits long, stronger algorithms are used compared to GSM. Nevertheless, privacy concerns remain. The services provided by EAP AKA include mutual authentication and the generation of a shared key for technology-specific security issues. Just as in EAP-SIM, identity protection cannot be guaranteed in all scenarios.

10.2.2.7 Other EAP Methods

Other EAP methods have not yet been discussed [11]. These are discussed briefly below.

EAP-message digest 5 (MD5) provides only user authentication by using user ID and password. It is vulnerable to dictionary attacks and man-in-the-middle attacks.

An EAP-secure remote password uses the DH method to authenticate both sides. The method provides mutual authentication and uses user ID and password.

EAP-SecureID uses OTP so as to authenticate the client. There is no authentication of the server in this method; the proposal is to use some sort of tunneling. Some issues can occur in this method, the main one being man-in-the-middle attacks.

References

[1]   IEEE 802.1X, "IEEE Standard for Local and Metropolitan Area Networks—Port-Based Network Access Control," July 2001.

[2]   Blunk, L., and J. Vollbrecht, "PPP Extensible Authentication Protocol (EAP)," RFC 2284, March 1998.

[3]   Asokan, N., V. Niemi, and K. Nyberg, "Man-in-the-Middle in Tunnelled Authentication Protocols," IACR ePrint Archive Report 2002/163, October 2002, http://eprint.iacr.org/2002/163.

[4]   Palekar, A., et al., "Protected EAP Protocol (PEAP) Version 2," Draft, draft-josefsson pppext-eap-tls-eap-07, IETF, October 2003.

[5]   Aboba, B., and D. Simon, "PPP EAP TLS Authentication Protocol," RFC 2716, October 1999.

[6]   Aboba, B., and M. Beadles, "The Network Access Identifier (NAI)," RFC 2486 (Standards Track), January 1999.

[7]   Funk, P., and S. Blake-Wilson, "EAP Tunneled TLS Authentication Protocol (EAP-TTLS)," Draft, draft-ietf-pppext-eap-ttls-03, IETF, August 2003.

[8]   Cam-Winget, N., et al., "EAP Flexible Authentication via Secure Tunneling (EAP-FAST)," Draft, draft-cam-winget-eap-fast-00.txt, February 2004.

[9]   Haverinen, H., and J. Salowey (eds.), "Extensible Authentication Protocol Method for GSM Subscriber Identity," Draft, draft-haverinen-pppext-eap-sim-13.txt, April 5, 2004.

[10] Arkko, J., and H. Haverinen, "Extensible Authentication Protocol Method for UMTS Authentication and Key Agreement (EAP-AKA)," Draft, draft-arkko-pppext-eap-aka-12.txt, April 5, 2004.

[11] EAP charter, http://www.ietf.org/html.charters/eap-charter.html, June 2, 2004.

11

WPAN Security

11.1 Introduction

Mobile platforms can form a wireless, personal, ad-hoc network whose security issues are becoming a common concern. Because of the physical limitations of these mobile platforms, such as limited computational ability and memory resource, frequent and unpredictable mobility, and strict power usage, conventional security technologies may not be as effective to achieve similar security goals as in other networks. In wireless personal ad-hoc networks, there is no fixed infrastructure, so it is difficult to establish a central authentication service, which means that common mechanisms cannot be applicable. The adaptability of security mechanisms becomes the key aspect in accomplishing security in personal ad-hoc wireless networks.

This chapter covers the security aspects of several WPAN protocols. In Section 11.2, Bluetooth security is described. Section 11.3 covers other emerging WPAN protocols, such as Zigbee and ultrawideband (UWB). More depth is provided for Bluetooth security because of the clear market penetration of Bluetooth devices, such as cellular phones. Market analysis firms estimate that about 500 million Bluetooth devices were shipped in 2006, up from about 30 million such devices shipped in 2005, and currently there are over 1.4 billion Bluetooth-enabled devices in operation [1].

11.2 Bluetooth

11.2.1 Bluetooth Overview

Bluetooth security has gained an unexpected media coverage [2–5] because of several security incidents. However, most of those Bluetooth exploits are simply annoyances rather than genuine threats to personal or corporate assets. In the following sections, details about the Bluetooth security architecture are provided, and then an analysis of the various Bluetooth security shortcomings and issues is provided.

Bluetooth is a radio frequency (RF) specification for short-range, point-to-multipoint voice communication and data transfer. Bluetooth aims at providing a low-cost, high-efficiency, and low-power solution to personal wireless communication system. The Bluetooth wireless communication specifications were developed by the Bluetooth Special Interest Group (SIG) formed in May 1998. The founding members were Ericsson, Intel, Nokia, Toshiba, and IBM. Version 1.0 of the specification was approved in the summer of 1999, and several releases followed after that [1]. Bluetooth can be used to connect any device to another device, and it can also be used to form ad hoc networks, called piconets, of up to eight devices. In these piconets, one of the Bluetooth devices acts as a master and the other devices are slaves. The master device sets the frequency-hopping behavior of the piconet. Several can be connected to each other to form a scatternet.

The Bluetooth specifications categorize devices into three classes based on their power usage. Class 1 devices have a transmission power up to 100 mW and a range up to 100 meters, class 2 devices have a transmission power of 1–2.5 mW and a 10-meter range, and class 3 devices have a 1 mW transmission power and a range of 0.1–10 meters [1, 6].

The main components of the Bluetooth architecture are the radio, the base frequency part, and the link manager protocol (LMP). Bluetooth uses the radio range of 2.45 GHz, has a theoretical maximum bandwidth of 1 Mbps, which can be greatly enhanced with the enhanced data rates (EDR) capability, and has a frequency hopping that's based on Gaussian frequency shift keying (GFSK). Bluetooth also has a base frequency that combines circuit and packet switching and can support data and speech channels. The LMP's main purposes are to configure, authenticate, and handle the connections between Bluetooth devices. Furthermore, the LMP operates the power management's three modes, which are sniff, hold, and park.

11.2.2 Bluetooth Security

11.2.2.1 Introduction

Bluetooth security consists of key management and generation mechanisms, a pairing protocol, authentication protocols, and encryption protocols [1, 7]. These protocols are implemented on the baseband layer, which is in turn controlled by upper layers, such as the link manager (LM), the logical link control, and the adaptation layer protocol (L2CAP).

The Bluetooth link-level security has four main components: the Bluetooth device address, the private authentication key, the private encryption key, and a random number. The device address is a 48-bit address that is unique for each Bluetooth device. The private authentication key is a 128-bit number that is utilized for authentication purposes. The private encryption key is 8-128 bits in length and is used for encryption purposes. The random number is 128 bits and is generated by the Bluetooth device. The Bluetooth passkey or PIN code has a length of 8-128 bits.

Bluetooth has different security modes and levels. The Bluetooth specifications define three modes: security mode 1(nonsecure), security mode 2 (service-level enforced security), and security mode 3 (link-level enforced security). In security mode 3, the Bluetooth device initiates security procedures before the channel is established. Furthermore, Bluetooth has different security levels for devices and services. For devices, there are two security levels: *trusted device* and *untrusted device*. This trust relationship is established during the pairing process, after which a device is verified as trusted when a positive authentication response is given and the trusted flag is set. For services, there are three security levels: services that require authorization and authentication, services that require authentication only, and services that are open to all devices.

11.2.2.2 Security Architecture

Above the link level, the Bluetooth security architecture includes several protocol and application layers, as shown in Figure 11.1.

The L2CAP and radio frequency communication (RFCOMM) protocols are Bluetooth-specific. L2CAP provides connectionless and connection-oriented data services to upper layer protocols with protocol multiplexing capability, segmentation and reassembly operation, and group abstractions. RFCOMM is a transport protocol for emulating RS-232 serial ports over the L2CAP protocol.

The LM uses the LMP to configure and handle connections between Bluetooth devices. Furthermore, the LMP manages Bluetooth's power management.

The service database of the security manager stores the device and service security information. In the absence of this registration, Bluetooth security defaults into required authorization and authentication for all incoming connections as well as required authentication for all outgoing connections.

The security manager decides whether the connection requests can be accepted and what kind of security functions can be applied to this connection using the security level of the devices and the requested services. The security manager performs the following six tasks: the initialization of the pairing process as well as the querying of the passkey entry by the user, the registration of the services provided on the device, the storage of the device's security-related information, answering the accesses requested by either protocols or applications, the enforcement of authentication or encryption before providing a connection to

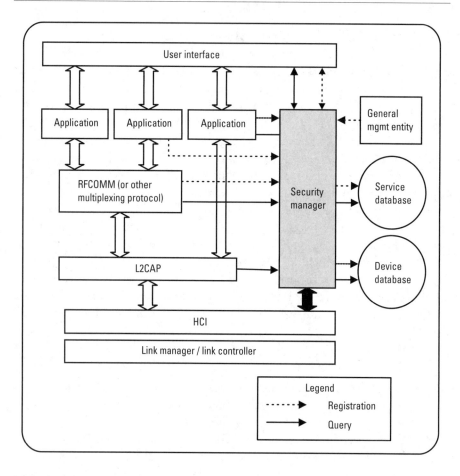

Figure 11.1 Bluetooth security architecture.

an application, and the initialization and processing of the inputs from a device's user in order to setup the necessary trust relationships.

11.2.2.3 Bluetooth Keys

Bluetooth defines several types of keys, which can be grouped in two categories: link keys and encryption key.

Link Keys   In Bluetooth, link keys handle all security transactions between devices. The link key is a 128-bit number and is used both in the authentication process and for deriving the encryption key. Link keys can be either semi- permanent or temporary. A semipermanent link key can be used after any specific session in order to authenticate Bluetooth devices that share it. A temporary link key's lifetime is one session and is typically used in point-to-multipoint connections, where the same information is transmitted to several recipients.

Also, a link key can be a combination key, a unit key, a master key, or an initialization key, depending on the type of application. A unit key is generated in a single device when it is installed. A combination key is derived from information from two devices, and it is generated for each new pair of Bluetooth devices. A master key is a temporary key, which replaces the current link key and can be used when the master unit wants to transmit information to more than one device. An initialization key is used as link key during the initialization process when unit or combination keys are not generated yet.

Initialization and Master Keys   The Bluetooth initialization key has a length of 128 bits and is used for devices that do not have prior relationships. The initialization key is generated using the *E22* algorithm as shown in Figure 11.2, which uses the following parameters as inputs: the passkey, the Bluetooth device address of the claimant device, and a 128-bit random number generated by the verifier device. The resulting initialization key is used for key exchange during the generation of a link key and is destroyed after initialization key exchange.

The Bluetooth master key is a temporary link key and is generated by the master device using the *E22* algorithm with two 128-bit random numbers, as shown in Figure 11.2. Once generated, the master key undergoes a bitwise XOR operation with the overlay and is then sent to the slave device. This device can then compute the master key.

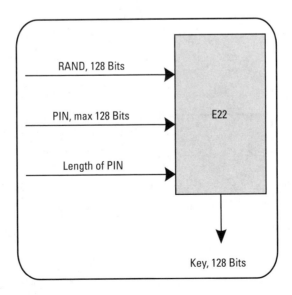

Figure 11.2 Master and initialization key generation.

Unit and Combination Keys   When a Bluetooth device is in operation for the first time, a unit key is generated with the key-generating algorithm *E21*, as shown in Figure 11.3. The unit key is then stored in the nonvolatile memory of the device, and another device can use it as a link key between the two devices. The application decides which device should provide its unit key as the link key during the initialization process, and the link key of the device with restricted memory capabilities is utilized.

Also, during the initialization process, a combination key may be generated by both devices at the same time if they decide to use one. After both devices generate a random number, they utilize the key-generating algorithm *E21* in order to generate the combination key by combining that random number and their Bluetooth device addresses, as shown in Figure 11.3.

Encryption Keys   Using the key-generation algorithm *E3*, an encryption key is produced from the current link key, a 96-bit ciphering offset number (COF), and a 128-bit random number, as shown in Figure 11.4. The COF is based on the ACO, which is generated during the authentication process. When the LM activates the encryption process, the encryption key is generated and is then automatically changed every time the Bluetooth device enters the encryption mode.

11.2.2.4 Bluetooth Pairing

When two Bluetooth devices that are creating a connection require an authentication, they first check if they already share a link key. In the case where they share a link key, they can initiate the authentication. In the case where they do not share such a key, they proceed to create an initialization key. The two

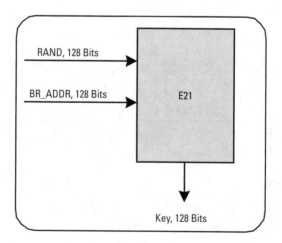

Figure 11.3 Unit and combination key generation.

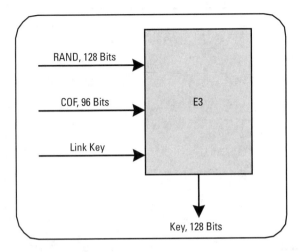

Figure 11.4 Encryption key generation.

devices decide whether the combination key or the unit key should be used as the link key. In case one device has limited memory resources, it can choose to utilize its unit key as the link key so that it does not have need to keep several keys.

For a device's first operation, it must generate its unit key from the initialization key, its Bluetooth device address, and a random number. However, in the case where the two devices want to share their own secrets, they can use a combination key as the link key. When a master device decides to broadcast to several slave devices, it can use a master key.

11.2.2.5 Bluetooth Authentication and Encryption

Bluetooth authentication uses a symmetric key-based challenge-response protocol in order to check whether the other device knows the secret key. During authentication, the ACO is generated and then stored on both devices. This ACO is used for generating the encryption key. The encryption function is described in Chapter 3.

The authentication scheme is shown in Figure 11.5. The verifier device first sends the claimant device a random number to be authenticated. Both devices then use the authentication function $E1$ with the random number, the claimant's Bluetooth device address, and the current link key to get a response. The claimant device then sends the response to the verifier device, which tests for matching responses. Authentication functions are described in Chapter 7.

In the case of failure of the authentication, new authentication attempts can only be made after a period of time, and this period of time doubles for each subsequent failed attempt from the same Bluetooth device address until the

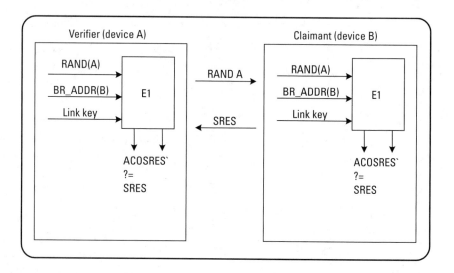

Figure 11.5 Bluetooth authentication.

maximum waiting time is reached. This waiting time will decrease exponentially in the case where no failed authentication attempts are made during a specific time period.

In the Bluetooth authentication scheme, the verifier device may not be the master device, and both one-way and mutual authentications are supported.

Bluetooth encryption is performed using a stream-cipher $E0$, which is resynchronized for every new payload. The Bluetooth encryption scheme is shown in Figure 11.6. The Bluetooth $E0$ stream-cipher is made up of three elements: a payload key generator, a key stream generator, and an encryption/decryption component. The payload key generator takes the input bits and then shifts them to the four LSFRs of the key-stream generator. The key stream bits are generated by a method derived from a summation stream cipher generator.

The Bluetooth encryption algorithm uses four LFSRs, with a total length of 128 bits. The initial 128-bit value of the four LFSRs is derived from the key-stream generator using the following four parameters as input: the encryption key, a 128-bit random number, the Bluetooth address of the device, and the 26-bit value of the master clock. The LFSRs use primitive feedback polynomials with a specified Hamming weight of five.

It is worth noting that, in Bluetooth, the specific size of the 8–128-bit encryption key is negotiated between the two devices. During key-size negotiation, the master device sends its proposed encryption key size to the slave device, which can either accept it or propose another size. This negotiation can either lead to a consensus or to an abortion of negotiations by one of the two devices.

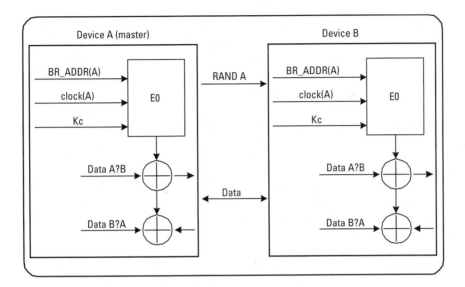

Figure 11.6. Bluetooth encryption.

There are several Bluetooth encryption modes, depending on whether a unit/combination key or a master key is used. If a unit key/combination key is used, broadcast traffic is not encrypted. However, if a master key is used, three encryption modes are available: encryption mode 1, encryption mode 2, and encryption mode 3. For the first mode, no packets are encrypted. For the second mode, broadcast traffic is not encrypted, but the individually addressed traffic is encrypted with the master key. And, for the third mode, all traffic is encrypted with the master key.

11.2.3 Bluetooth Security Challenges and Recommendations

Bluetooth security gained wide media coverage because of several Bluejacking incidents [2–5]. However, those Bluetooth exploits are merely annoyances at this point and have not been used as security attacks where personal or corporate assets were jeopardized or services were illegally used.

The following is a list of known Bluetooth attacks:

- *Bluejacking:* Bluejacking allows Bluetooth-enabled mobile phone users to send business cards anonymously over a Bluetooth connection. Once the receiving individual accepts to receive the message, she will get a more personal message on her Bluetooth device.

- *Bluebugging and Bluesnarfing:* These vulnerabilities are more serious than Bluejacking. Bluebugging allows an attacker to access the mobile phone commands using Bluetooth without alerting the phone's user. Then, this attacker can initiate phone calls, send and receive text messages, read and write phonebook contacts, eavesdrop on phone conversations, and connect to the Internet. Bluesnarfing allows an attacker to gain access to data stored on a Bluetooth-enabled phone without alerting the user of the connection. The data that can be accessed can include the user's phonebook, calendar, and IMEI.

- *DoS attacks:* A constant request for response from an attacker's Bluetooth-enabled device to a victim's Bluetooth-enabled device can result in battery degradation in the victim's device or can even disable that device.

- *Backdoor attack:* The backdoor attack requires that a trust relationship has been previously established between two devices. Following this, the pairing relationship must be removed from the victim device's listing of paired devices.

- *Snarf attack:* This attack effectively sidesteps the need for devices to be paired in order to navigate a device's data. This attack is much easier to perform than the backdoor attack, as there is no need to even have contact with the victim. The victim is never notified of requested connection attempts, and an attacker can easily gain access to contacts, phone logs, and other sensitive information. This attack is far more likely to happen in a real-world scenario than the backdoor attack as it does not require devices be paired in order to perform it. There are ways to mitigate the risk of data exposure, such as placing the device in nondiscoverable mode.

More serious, but still more complex attacks on Bluetooth include:

- Pairing attacks, where an attacker attempts online or offline PIN recovery.

- Cipher attack, where an attacker tries to break the $E0$ algorithm.

It is agreed that, although sufficient for simple usages like phone-headset connection, current Bluetooth security suffers several weakness that do not allow for safe implementation of emerging usages like high-value m-commerce. The main shortcomings in the current Bluetooth security model are:

- The encryption keys are derived from the link keys, which in turn are derived from the passkey (PIN code). Most users enter small PINs, resulting in weak encryption, as a four-digit PIN code results in only 10,000 different possibilities.

- The $E0$ stream cipher with 128-bit key length can be relatively easy to break. The Bluetooth encryption and key-derivation algorithms have not been improved, although several stronger algorithms have been available. Also, the Bluetooth architecture does not permit for cipher negotiation (as SSL and TSL do), so there is no provision for using stronger ciphering algorithm.

- The default Bluetooth security has no authentication or encryption.

- Use of the unit key is not recommended.

Many researchers have come up with new methods to address the Bluetooth security limitations. Enhancements to the pairing mechanism, the key exchange protocol, and the cipher algorithm have been proposed by several researchers [8, 9]; however, with the upcoming merger of Bluetooth and UWB, it is expected that improvements to the security model will be specified.

A relatively stronger security in Bluetooth can be achieved today through upper layer protocols or applications with strong policies. The Bluetooth security architecture does not prevent the enforcement of the security policies of upper applications. Enforcement of stronger PINs and higher-trust modes can also improve the Bluetooth security. Furthermore, a SSL/TLS tunnel can be established on top of the Bluetooth connection for high-value transactions.

11.3 Other Emerging WPAN Technologies

This section briefly describes the two most promising emerging WPAN technologies, which are still being defined at the writing of this book.

11.3.1 Zigbee

11.3.1.1 Introduction

The Zigbee specifications [10] were developed by the Zigbee Alliance, an association of companies involved with developing higher-layer standards based on IEEE 802.15.4. Besides local device connections, Zigbee is also applies to sensor networks.

In Zigbee, a data packet is of variable length and can be used for unicast or broadcast. Every data packet has a two-byte flag field that indicates whether or not security is set, the addressing mode that is used, and whether the sender requests acknowledgment from the receiver.

The identification of the packet number for acknowledgment is achieved using the sequence number, which also includes the destination and source addresses. The payload is also of variable size up to 102 bytes. There is also a two-byte cyclic redundancy check (CRC) checksum for error correction.

An acknowledge packet is sent when the sender requests acknowledgment, and the packet is not broadcasted. This acknowledge packet has similar length and flags bytes as a data packet.

11.3.1.2 Security Aspects

The Zigbee architecture [10] includes several security aspects, such as access control, message integrity, message confidentiality, and replay protection [11, 12].

For message authentication and integrity, a sender over a Zigbee network calculates the MIC of the message using a shared secret key, appends the MIC to the message, and then transmits both. The receiver, who shares the secret key with the sender, recalculates the MIC and compares it with the MIC that came with the message. Only when both MICs match is the message accepted as authentic. The MIC has to be hard to forge without knowledge of the secret key.

In Zigbee, confidentiality is achieved using a unique nonce in the encryption algorithm. This nonce adds variation to the encryption process.

For replay protection, the sender device assigns a monotonically increasing sequence number to each packet, and the receiving device detects and stores that sequence number. Receiving a packet with an old number means a replay attack.

Zigbee includes several security suites, which are supported using access control lists (ACLs). Zigbee defines up to 255 ACL entries. When an applications needs to communicate using security services with another device, the transmitting device looks for the address of the destination device in the ACL, and if finds it, it will use the security suite, key, IV, and replay counter to communicate securely. The receiving device checks the flag fields to determine if any security suite has been applied to the packet.

The Zigbee specifications for group keying are not very specific. A network with several Zigbee-enabled nodes requiring confidential services can be implemented with several ACLs, all of them with the same key. In this case, the nonces can be repeated, and the security of the connection can be compromised.

Furthermore, the Zigbee specifications for low-power or power-failure situations are not very specific. In the case of power failure, the ACL state may be lost and counters may be reinitialized. This may result in nonces being repeated, which can compromise the security of the network. Also, when a Zigbee-enabled device transitions to a low-power state, nonces could be reused in case the state of the ACL entries is not preserved.

11.3.2 Ultrawideband

11.3.2.1 Introduction

Ultrawideband (UWB) originates from the impulse radio work in the early 1960s. UWB is a radio technology with a spectrum that occupies greater than 20 percent of the center frequency or a minimum of 500 MHz. In 2002, the Federal Communications Commission (FCC) allocated unlicensed radio spectrum from 3.1 GHz to 10.6 GHz merely for UWB purposes, and more spectrums are available for other usages, such as law enforcement and medical emergency. The FCC also defined a minimum bandwidth of 500 MHz at a −10-dB level. UWB-enabled devices can maintain the same low transmit power as if they were using the entire bandwidth by interleaving the symbols across these subbands. Information can either be transmitted by the traditional pulse-based single carrier method or by more advanced multicarrier techniques.

UWB uses modulation techniques, such as orthogonal frequency division multiplexing (OFDM), to occupy extremely wide bandwidths. The multiband OFDM (MB-OFDM) provides very good coexistence with narrowband systems such as 802.11a, adaptation to various regulatory environments, and future scalability and backward compatibility. MB-OFDM transmits data simultaneously over multiple carriers spaced apart at precise frequencies. MB-OFDM provides high spectral flexibility and resiliency to RF interference and multipath effects. In MB-OFDM, the available spectrum of 7.5 GHz is divided into several 528-MHz bands, which allows the selective implementation of bands at certain frequency ranges, while leaving other parts of the spectrum unused.

Furthermore, the information transmitted on each band is modulated using OFDM. OFDM distributes the data over a large number of carriers that are spaced apart at precise frequencies, which provides sufficient orthogonality and prevents the demodulators from seeing frequencies other than their own. Hence, OFDM can provide high spectral efficiency, resiliency to RF interference, and lower multipath distortion.

11.3.2.2 Security Aspects

For UWB, the security mechanisms are implemented at several levels of the protocol stack [13]. Because of their low average transmission power, UWB communications systems have an inherent immunity to detection and interception. Such low transmission power requires an eavesdropper to be very close to the transmitter (about 3 ft) to be able to detect the transmitted information.

Since UWB pulses are time modulated with codes that are unique to each transmitter-receiver pair, this time modulation of very narrow pulses adds more security to the UWB transmission. Detecting pico-second pulses without knowing when they would arrive is a very difficult undertaking. This time modulation can achieve a low-probability of interception and detection by an attacker.

References

[1]   Bluetooth Specifications, Bluetooth SIG, http://www.bluetooth.com.

[2]   http://www.eweek.com/article2/0,1759,1526640,00.asp.

[3]   http://news.zdnet.co.uk/0,39020330,39145886,00.htm.

[4]   http://www.wired.com/news/privacy/0,1848,64463,00.html.

[5]   http://www.computerworld.com/mobiletopics/mobile/story/0,10801,93031,00.html.

[6]   Bray, J., *Bluetooth Application Developer's Guide,* Sebastopol, CA: Syngress, 2001.

[7]   Müller, T., "Bluetooth Security Architecture," http://www.bluetooth.com/Bluetooth/ Apply/Technology/Research/Bluetooth_Security_Architecture.htm.

[8]   Aissi, S., C. Gehrmann, and K. Nyberg, "Proposal for Enhancing Bluetooth Security Using an Improved Pairing Mechanism," http://www.3gpp.org/ftp/TSG_SA/WG3_Security/ TSGS3_34_Acapulco/Docs/PDF/S3-040481.pdf.

[9]   Gehrmann, C., J. Persson, and B. Smeets, *Bluetooth Security,* Norwood, MA: Artech House, 2004.

[10]   Zigbee Specifications, http://www.zigbee.org/en/spec_download/download_request.asp.

[11]   Sastry, D., and N. Wagner, "Security Considerations for IEEE 802.15.4 Networks," *Proceedings of the 2004 ACM Workshop on Wireless Security,* October 2004.

[12]   Vines, R. D., *Wireless Security Essentials,* New York: Wiley Publishing, 2002.

[13]   "Ultra-Wideband Specifications," http://grouper.ieee.org/groups/802/15/pub/2003/ Jul03/03268r2P802-15_TG3a-Multi-band-CFP-Document.pdf.

12

WLANs Security

WLANs, particularly IEEE 802.11, have gained much more ground than was expected within a very short period of time [1–3]. This growth is being hampered by security issues [4–24]; the issue is by now so well publicized that even laypeople are aware of it.

Wireless communication networks encounter greater security issues than do wired ones because of their intrinsic characteristics. Attacks can be performed in a wireless medium remotely and over the air. An attacker does not need to gain physical access to the wires that allow data transmission; she only needs her receiver and/or transmitter to be within reach of the wireless communication station. Also, the attack may not aim at a single person but can be repeated toward all users within reach. For these reasons, standardization bodies make efforts to secure the communication at the MAC level, whereas it is generally accepted that data over wired communication lines travels in clear. Moreover, low-level security is enforced even though application-level security or VPNs are also implemented. As stated in the name for WLAN encryption, wired equivalent protocol, the goal for low-level security in wireless communication networks is to achieve a similar level of trust as the one that is put in wired communication means.

This chapter starts with a short background on security, after which some security protocols are discussed. Following that, the security issues in the IEEE 802.11 countermeasures present in the market are discussed. Wi-Fi protected access (WPA) and the IEEE 802.11i standard (ratified in June 2004) are explained next. Finally secure WLAN deployment for corporate, public, and mobile operators are given.

12.1 Security in IEEE 802.11

The original IEEE 802.11 standard provided authentication, integrity, and confidentiality. The standard did not provide any key exchange mechanism but allowed higher layer solutions to be used. The security solution in the original standard is WEP. At the time when WEP was designed (about 15 years ago), it was considered reasonably secure; one can understand the reason too: Netscape, Microsoft, Oracle, and Lotus provided similar security levels.

In this section, the original security solution of IEEE 802.11 is explained. The section also discusses the security issues in WEP.

12.1.1 Authentication

IEEE 802.11 defines two subtypes of authentication service: open system and shared key [1, 2].

Open system authentication is the simplest of the available authentication algorithms. Essentially, it is a null authentication algorithm. Any station that requests authentication with this algorithm may become authenticated if the recipient station is set to open system authentication (see Figure 12.1).

Shared key authentication supports authentication of stations as either a member of those who know a shared secret key or a member of those who do not. IEEE 802.11 shared key authentication accomplishes this without the need to transmit the secret key in the clear, requiring the use of the WEP mechanism. Therefore, this authentication scheme is only available if the WEP option is implemented. The required secret shared key is presumed to have been delivered to participating stations via a secure channel that is independent of IEEE 802.11. During the shared key authentication exchange, both the challenge and the encrypted challenge are transmitted. This facilitates unauthorized discovery of the PRN sequence for the key/IV pair used for the exchange. Therefore the

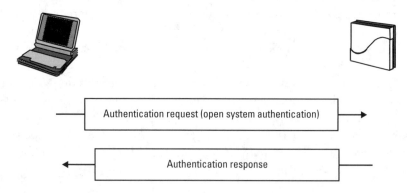

Figure 12.1 Open system authentication.

same key/IV pair for subsequent frames should not be used. The shared key authentication process is given in Figure 12.2.

12.1.2 Encryption

The WEP algorithm is a form of electronic code book in which a block of plaintext is bitwise XORed with a pseudorandom key sequence of equal length. The key sequence is generated by the WEP algorithm.

Referring to Figure 12.3 and viewing from left to right, encipherment begins with a secret key that has been distributed to cooperating stations by an external key management service. WEP is a symmetric algorithm in which the same key is used for encipherment and decipherment.

The secret key is concatenated with an IV, and the resulting seed is input to a PRNG. The PRNG outputs a key sequence k of pseudorandom octets equal

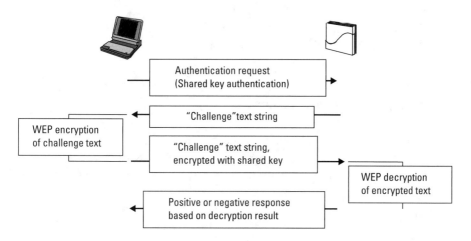

Figure 12.2 Shared key authentication.

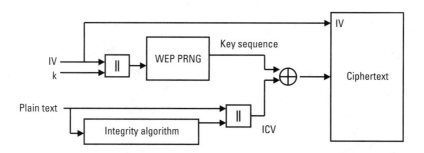

Figure 12.3 WEP encipherment block diagram. (*Source:* IEEE 802.11i, reprinted with permission from IEEE.)

in length to the number of data octets that are to be transmitted in the expanded MAC protocol data unit (MPDU) plus 4 (since the key sequence is used to protect the ICV as well as the data). Two processes are applied to the plaintext MPDU. To protect against unauthorized data modification, an integrity algorithm operates on P to produce an ICV. Encipherment is then accomplished by mathematically combining the key sequence with the plaintext concatenated with the ICV. The output of the process is a message containing the IV and ciphertext.

Referring to Figure 12.4 and viewing from left to right, decipherment begins with the arrival of a message. The IV of the incoming message shall be used to generate the key sequence necessary to decipher the incoming message.

Combining the ciphertext with the proper key sequence yields the original plaintext and ICV. Correct decipherment shall be verified by performing the integrity check algorithm on the recovered plaintext and comparing the output ICV′ to the ICV transmitted with the message. If ICV′ is not equal to ICV, the received MPDU is in error and an error indication is sent to MAC management. MSDUs with erroneous MPDUs (due to inability to decrypt) shall not be passed to LLC. The WEP payload is shown in Figure 12.5.

12.2 IEEE 802.11 Security Issues

Although the IEEE 802.11 standard-based WLAN has evolved into a major wireless technology, its growth is being hampered by security flaws. In this section the security issues in the current IEEE 802.11 standard are discussed; the

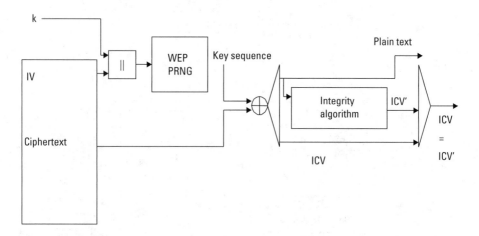

Figure 12.4 WEP decipherment block diagram. (Source: IEEE 802.11i, reprinted with permission from IEEE)

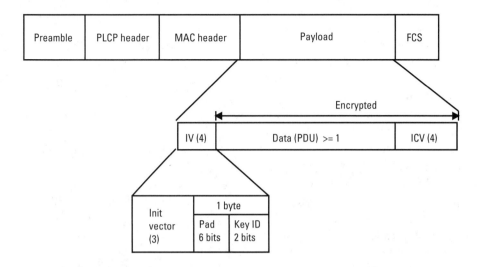

Figure 12.5 WEP payload [1] (Reprinted with permission from IEEE 802.11, 1999). (From: IEEE std. 802.11i—2004. Copyright 2004 IEEE. All rights reserved.)

security solution itself is discussed in Section 3.4. The authors have tried to present the security issues from the point of view of security goals so as to map them with the discussion in Section 12.1. Availability and non-repudiation are not considered because availability is more related to the network and not to the protocol, while non-repudiation is related to the service. Note that there are several security issues concerning IEEE 802.11 that appear every day; thus, the list might not be complete. The WEP algorithm has been shown to have several weaknesses [2, 5–15].

12.2.1 Authentication

IEEE 802.11 provides two types of authentication mechanisms; these are open system and shared key-based. The first method is obviously a *nonsecurity* solution; everyone is let in. The second method is dependent on *shared key*; that is, each station of the network has the same key, and so does the access point (AP). Thus the AP authenticates the station.

There are flaws in key generation and key management, discussed in a later section, that affect the authentication. Note that this is a very basic security architecture flaw; one should separate the authentication and encryption (confidentiality) keys. The flaw from the authentication point of view is that the shared key is used for a very long time, as renewing the key in each station means a lot of overhead for the IT staff [2].

Further, the use of challenge and response makes it easier for an attacker to break in. As discussed in Section 3.4., the data is XORed with the key sequence to produce an encrypted message. This means the attacker just has to capture the challenge or plaintext and the encrypted response and then XOR the two giving the key sequence. Now the attacker can simply request authentication, encrypt the challenge with the key sequence, and send it to the AP. The AP will simply decrypt and authenticate the station [2].

Another authentication issue is that it is usually not per station–based, although key per MAC address is possible; neither is the key different per AP. Once again this issue is related not only to the standard, but also to the IT overhead [2].

Even if one could provide the station-based authentication, it remains at the air interface level (i.e., the AP and the station). The point is to provide per user authentication, which is not possible with the current IEEE 802.11 solution and is understandable because it is a layer 2 solution. Thus, it is important to integrate a higher layer authentication mechanism [2].

An important issue is related to mutual authentication. IEEE 802.11 does not provide mutual authentication; thus, an AP can verify a station but not the other way around [2]. This is an important issue because the station can easily attach itself to a rogue AP or network.

With the shared key mechanism, let us not forget the issue of stolen stations and the time taken to discover that a station is stolen to informing the IT and renewing the secret key [2].

12.2.2 Confidentiality

Usually one would say confidentiality, which is provided by encryption, is dependent on authentication (i.e., once authentication is successful, confidentiality can be provided). This does not mean that the keys should be the same for confidentiality and authentication. For a WLAN network, knowing that authentication has its share of weakness one could simply skip the authentication of station at the air interface level and use confidentiality to provide user authentication at the network level. Thus one is talking about confidentiality and then authentication.

Encryption and integrity issues are discussed in Chapter 3; here we will discuss the issues related to key generation and key management as alluded to in the previous section.

12.2.2.1 Key Management

There is no real key management in WEP, but two methods of using WEP keys are provided [2]. The AP and the stations share the usage of the four (default) keys. The compromising of each of the nodes means a compromise of the wireless network. A key mapping's table is used at the AP. In this method, each unique MAC address can have a separate key. The size of a key mapping's table should be at least 10 entries, according to the 802.11 specification; however, it is likely chip-set dependent. The use of a separate key for each user mitigates the known cryptographic attacks but requires more efforts on the manual key management. Since key distribution is not defined in the WEP and can be done only manually, many of the organizations deploying wireless networks use either a permanent fixed cryptographic variable, or key, or no encryption at all.

12.2.3 Access Control

Let us look at this goal in two parts, the first being the access control itself and the second being authorization.

12.2.3.1 Access Control

Although IEEE 802.11 does not define access control, it can be provided by using MAC address. The AP can have a list (ACL) of MAC addresses that are allowed to access the network. It is relatively easy to forge MAC addresses, so this is not really a secure solution; still it is used by several organizations.

12.2.3.2 Replay Attack (Authorization)

IEEE 802.11 does not provide any form of replay protection, which also helps in other security attacks. Any intercepted message can be simply resent. There is no concept of timestamp and packet numbering in the security mechanism. This lack of authentication can permit attackers to launch potential man-in-the-middle attacks or DoS attacks.

12.2.4 Other Issues

In this section some other security issues related to IEEE 802.11 are discussed.

12.2.4.1 Password Protection

Most APs being sold today have a password to access the management functionalities. Very often it is noted that these passwords are not changed due to ease of use. Thus an attacker can simply try to use the default product password of various well-known vendors and modify the settings of the AP for its benefit. Included in this is the issue of the factory default state or reset method in APs. These problems are not only for WLANs but also for network elements in mobile networks or in fact any system using default passwords.

12.2.4.2 Location of APs

Location of APs should be such that they are not stolen easily, or else there should be an alarm on the network or the device. A stolen AP can be used as well as a stolen station.

12.2.4.3 DoS Attack

A DoS attack can be performed in a wireless medium very simply by creating a lot of interference, but let us look at some specific situations. Within IEEE 802.11 there is no integrity protection for management frames; this means that an attacker can simply send a dissociation or deauthentication to the station or the AP, not only that a broadcast MAC address can also be used to disconnect all the stations (STAs) connected to the AP.

An attacker can also use the network allocation vector (NAV) field and set a long NAV time to create DoS. The NAV field is used to indicate other devices in the network the period during which a device will transmit so that other devices do not try to access the channel in that period.

12.2.4.4 Man-in-the-Middle Attack

Man-in-the-middle attacks are also possible using some of the thoughts presented in the previous section. The attacker can simply disconnect a user and act as a fake AP. In this case, the station will try to connect to the fake AP (lack of mutual authentication); the attacker will then simply pass the connection information to the AP and thus will be in a situation to do several things, including modification of frames and accessing the network.

It is also possible to give fake a response to an ARP request and masquerade as a user. Thus, the traffic of a user will go to the attacker.

12.2.4.5 DHCP

The DHCP provides an automatic IP address to the stations from an address pool. There is usually no DHCP authentication; thus, a stolen device or cracked network will give direct access to the network.

12.2.4.6 Management

The simple network management protocol (SNMP) is often used for network management and monitoring. It should be noted that SNMPv1 and SNMPv2 do not have much security support, and this opens a path for the attacker. It is normally advised to use SNMPv3 with better security support.

12.2.5 Tools

There are several tools available on the Internet today that can be used for attack purposes, but there is also a positive side. These tools can also be used to find security issues in the WLAN network and even sometimes for network planning purposes. Several such tools are listed in [16].

One of the most famous tools for *war driving* is NetStumbler [17]. War driving basically maps the APs that are available in a given region with no security. It also comes as MinioStumbler for PDAs. Another tool that works on Linux is Kismet, with all the functionalities of NetStumbler [18].

For just sniffing purposes, one could use the tools like Ethereal [19], which works for both Windows and Linux. There are several others, some of which are listed under [16].

Tools for forging MAC addresses are SMAC for Windows and MAC change for Linux [20, 21]. The tools that can be used for cracking WEP are AirSnort and WEPCrack; both are for Linux [22, 23].

If one is not satisfied with just cracking the WEP, then there are also tools to perform man-in-the-middle attacks. The most well-known one is AirJack [24].

12.2.6 Security Issues in Other Solutions

In this section, security issues related to IEEE 802.1X, PEAP, and TTLS are discussed briefly.

It was found in [15] that IEEE 802.1X has a major flaw. Basically man-in-the-middle attacks are possible when IEEE 802.1X is used. The attacker waits until the station is authenticated and then sends a disassociation or deauthentication message to the station. For the AP, the session is still alive; thus, the attacker can now use the session until reauthentication. The man-in-the-middle attack method presented in Section 12.2.4.4 can be used by the attacker too.

In the case of PEAP or TTLS, there is the possibility of the attacker masquerading as a station to the AP and as an AP to the station. The attacker creates a tunnel with the AP and the station; the first phase is simple as anyone can get in as an anonymous person. Next, the attacker simply passes the station's responses to the AP, and, on achieving the connection with the AP, simply disconnects the station.

12.3 Countermeasures

There are several countermeasures for the security issues in WLANs. In this section let us look at them briefly. The following section will give further information on a few higher layer protocols that can be used.

Some of the methods to provide countermeasure are using personal firewalls, intrusion detection systems (IDSs), VPNs [25], PKI [26], and possibly biometrics. Note that these are in addition to the correct methods of configuring APs and WLANs.

It is advisable to install personal firewalls on user devices whether they are from a company IT department or personal.

Biometrics is a method of identifying users based on their physical information (e.g., fingerprints or reading the retina [25]). Although biometrics is an option, it should be noted that it is not a perfect science. Human physical condition changes with time; further, solutions are also not yet available that can work independently. It is better to use biometrics, in its current stage, with another security solution (e.g., biometrics can combine with VPN solutions).

WLANs can integrate PKI for authentication and secure network transactions. Third-party manufacturers, for instance, provide wireless PKI, handsets, and smart cards that integrate with WLANs. Smart cards provide even greater utility since the certificates are integrated into the card. Smart cards serve both as a token and a secure (tamper-resistant) means for storing cryptographic credentials.

CableLabs has paved the path toward developing a PKI solution for a complete industry (http://www.cablelabs.com). All manufacturers and thus their products have certificates, and so there is certification for the ISPs. This method when used by WLANs can allow user, device, and network authentication along with the other benefits of using certificates/PKI.

The IDS can be used as a host-based or network-based system or as a hybrid of the two systems. A host-based agent is installed in individual systems, while a network-based IDS monitors the network traffic. The IDS solutions are wired-specific and need to be modified to work in wireless medium. Some of the enhancements required are [27] determination of location of devices, detecting a rogue AP, and detection of attacks at the wireless medium between stations.

12.4 WPA and IEEE 802.11i RSN

While the security issues of IEEE 802.11 were widespread and the standard was still working on security enhancements or IEEE 802.11i, the WiFi Alliance came up with an intermediary solution known as WPA. For most existing products, a firmware upgrade can be used to provide WPA, but IEEE 802.11i requires a hardware upgrade. WPA makes use of TKIP and it is an optional mode in IEEE 802.11i [1, 2, 28–31].

IEEE 802.11i defines a robust security network (RSN) that requires a number of capabilities at the station and the AP [31]. The standard also defines a transitional security network (TSN) in which both WEP and RSN can work. Unlike WPA, RSN makes use of an AES.

The biggest differences between the original IEEE 802.11 security and WPA and RSN are the possibility of using higher layer protocols for authentication and the possibility of key exchange. Both of them make use of IEEE 802.1X, EAP, and other methods explained in Chapter 10.

A variety of keys are used in both WPA and RSN; this is first explained in this section. After the hierarchy is clarified, the TKIP and AES methods are explained. However, before all that, the benefits of WPA and RSN are given.

12.4.1 IEEE 802.11i Services

IEEE 802.11i provides the following services:

- *Association and reassociation:* An IEEE 802.1X port maps to an association. Once the STA is authenticated by a higher layer protocol and authorized to access the network, the AP allows data traffic for the STA.

- *Access control:* This is provided by use of IEEE 802.1X and higher layer protocols.

- *Authentication and deauthentication:* IEEE 802.11 provides link layer authentication using open system and shared key authentication. In IEEE 802.11i, authentication is provided by using IEEE 802.1X with EAP or by preshared key (PSK). When using IEEE 802.11i, the STA must first use open system authentication to perform authentication with the AP. (Note that shared key authentication is not allowed.) After open system authentication, the association takes place and then the IEEE 802.1X process starts. Deauthentication will terminate any association between the AP and the STA. Any authenticated party can send the deauthentication message, and this message cannot be refused, as it is not a request but a notification.

- *Confidentiality:* Confidentiality is provided by using WEP, TKIP, and counter mode with the CCMP. CCMP makes use of AES and the rest use RC4. The default confidentiality state is to send data in clear (i.e., not use any protection).

- *Key management:* All the new services in IEEE 802.11 require key management, which uses four-way handshake and group key handshake mechanisms to provide fresh keys.

- *Data origin authenticity:* Data origin authenticity is provided by TKIP and CCMP and is applicable to unicast frames only.

- *Replay protection:* Again, TKIP and CCMP provide replay protection.

12.4.2 RSN Information Elements

The RSN information element (IE) is sent in the beacon and probe response from the AP. The IE contains security information such as the kind of authentication supported, the cipher suite supported, and the key exchange mechanism supported. This is to be negotiated between the two parties.

On receiving the IE from the AP, the STA also sends an IE, which contains only a single choice of the different types of algorithms. This choice made by STA is used for further communication.

The IE is also sent during four-way handshake so as to prevent a bidding attack. The contents of IE can be changed during the handshake.

12.4.3 Key Hierarchy

IEEE 802.11i defines a pairwise key hierarchy for unicast traffic and a group key hierarchy for multicast and broadcast traffic. The basic element in the hierarchy is the pairwise master key (PMK). In the following, a short discussion is given on PMK, after which the two hierarchies are discussed.

12.4.3.1 PMK

PMK is the top of the key hierarchy. A pairwise shared key (PSK) can also be used as a PMK. In case PSK is not available, upper layer authentications methods (with the help of IEEE 802.1X and EAP) are used to create PMK. PMK is created from the AAA key, which is also sometimes known as the master key (MK). The AAA key is jointly negotiated between the STA and the AS. This key information is transported via a secure channel from the AS to the authenticator. The PMK is computed as the first 256 bits of the AAA key.

Once PMK is created between the STA and the AS by using IEEE 802.1X, it has to be transferred to the AP. IEEE 802.11i does not define any secure method for doing so, but it is defined in WPA.

12.4.3.2 Pairwise Key Hierarchy

Before starting a discussion of the key hierarchy, there are two functions used by IEEE 802.11i. These are the following:

- L(Str,F,L): From Str, starting from left to right, extract bits F through F+L-1.
- PRF-n: PRF produces n bits of output. PRF is a function that hashes various inputs to derive a pseudorandom value (the key).

The PMK is used to create pairwise transient keys (PTKs). Transient keys are used for confidentiality algorithms, and their maximum lifetime is PMK lifetime. PTKs are created for each association. The PTK consists of EAPOL-key confirmation key (KCK), the EAPOL-key encryption key (KEK), and temporal keys (TKs) for TKIP and CCMP. In the following list, these keys and their use are explained (see Figure 12.6, where the AA is the authenticator address and the SPA is the supplicant address). The nonces are explained in Section 12.4.3.4.

- KCK: It is 128 bits and is used by IEEE 802.1X in a four-way handshake (see Section 4.5.3.4) for data origin authenticity. One can also call this key the integrity key.

- KEK: This is also 128 bits long and is used in the handshake to provide confidentiality.

- TKs for TKIP: TKIP makes use of RC4, which only had the possibility for encryption. Thus TKs in TKIP consist of the integrity and encryption keys of 128 bits each. Bits 0–127 of TKs are input to the TKIP phase 1 and 2 mixing functions (i.e., for encryption) (see Chapter 3). Bits 128–191 of TK are used as the Michael key—that is, integrity key for MAC service data units (MSDUs) from the authenticator (AP) to the supplicant (STA)—while bits 192–255 are used as the Michael key for MSDUs from the STA to the AP. Note that a MSDU is a packet of data between the software and the MAC in contrast to the MPDUs, which are the MAC layer packets. Thus a MPDU can be a portion of the MSDU if the MSDU is bigger than MPDU.

- TKs for CCMP: In case of CCMP both encryption and integrity are incorporated in a single calculation. Thus there is one key of length 128 bits.

- KeyID 0: This is used when sending a pairwise key.

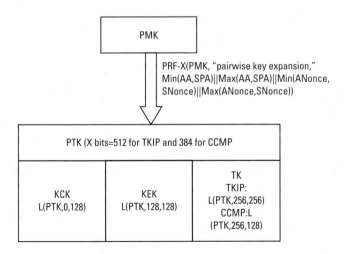

Figure 12.6 Pairwise key hierarchy (From IEEE std. 802.11i—2002. Copyright 2004 IEEE. All rights reserved.)

12.4.3.3 Group Key Hierarchy

A 256-bit group master key (GMK) is created. From this GMK, the group transient key (GTK) is created, out of which the group in the pairwise connection is established, and the GTK is sent to the STA, where the acknowledgment is checked. Similar to pairwise key hierarchy, TKIP uses one 128-bit key for encryption and one for integrity, while CCMP has one key for both purposes. The group key hierarchy is shown in Figure 12.7.

Group keys use the key rotation method. If a given group key is using KeyID1, then the new key is stored at KeyID 2. The new key is used as soon as keys in all STAs are updated.

A STA shall use bits 0–127 of TK as the input to the TKIP phase 1 and 2 mixing functions and bits 128–191 as the Michael key for MSDUs from the AP to the STA. Bits 192-255 of TK are used as the Michael key for MSDUs from the STA to the AP. Bits 0–39 and bits 0–103 are used as WEP-40 and WEP-104 keys, respectively. For CCMP, the TK is used as the key.

12.4.3.4 Liveness

As per IEEE 802.11i, liveness is a method to demonstrate that the peer is actually participating in this instance of communication. Thus its purpose is to prevent a replay of the same message in different sessions. Liveness is added to the PRF in the form of a nonce as one of the inputs. A nonce is guaranteed never to be reused.

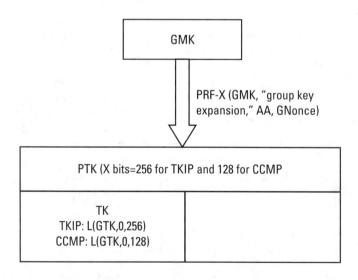

Figure 12.7 Group key hierarchy (From: IEEE std. 802.11i—2004. Copyright 2004 IEEE. All rights reserved.)

Each device generates a nonce and sends it to the other device. Both nonces are taken together with the two MAC addresses and the PMK to produce PTKs. The nonce from the authenticator (AP in our case) is called ANonce and from the supplicant (STA) is SNonce.

12.4.4 Handshake Protocols

After the PMK is transferred by the AS to the AP, the four-way handshake takes place to create the PTKs, and the two-way handshake takes place to create the GTKs. Both of these procedures are shown Figures 12.8 and 12.9, respectively.

12.4.5 SAs in RSN Association

Within RSN the STAs, the AP, and the AS create an association known as the RSN association (RSNA) by using IEEE 802.1X. Within RSNA secure communication takes place by using SAs. A SA is a relation between two communicating ends, which defines the method in which the secure communication will

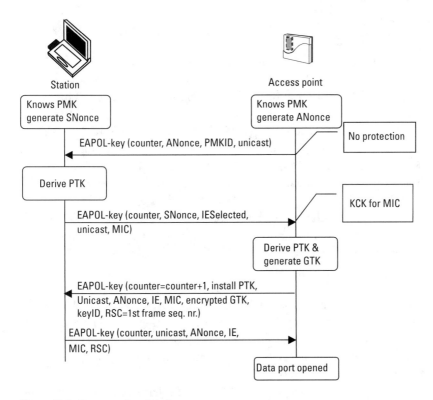

Figure 12.8 Four-way handshake.

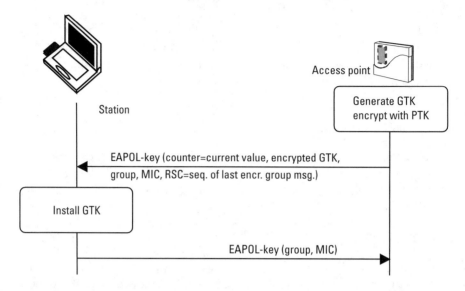

Figure 12.9 Two-way handshake.

take place; it is stored at both ends and contains an ID. There are four types of SAs defined in RSNA.

The result of a successful IEEE 802.1X authentication leads to a PMK Security Association (PMKSA) between an AP (authenticator) and a STA derived from EAP authentication and authorization parameters. This SA is bidirectional. The PMKSA is used to create the PTK Security Association (PTKSA). PMKSAs are cached for up to their lifetimes. The PMKSA consists of the PMK Identity (PMKID), which identifies the SA, AP MAC address, PMK (or PSK), lifetime, authentication and key management protocol (AKMP), and authorization parameters specified by AS or by local configuration at the AP.

The PTKSA is a result of the four-way handshake and is bidirectional. PTKSAs are cached for the life of the PMKSA. There is only one PTKSA with the same supplicant and authenticator MAC addresses. The PTKSA consists of the PTK, pairwise cipher suite selector, STA MAC address, and AP MAC address.

The GTK Security Association (GTKSA) results from a successful four-way or two-way handshake and is unidirectional. A GTKSA is used for encrypting and decrypting broadcast and multicast messages. A GTKSA consists of the direction vector (whether the GTK is used for transmit or receive), group cipher suite selector, GTK, AP MAC address, and authorization parameter specified by local configurations in the AP GTKSA.

The STAKeySA is a result of the STAKey handshake. This security association is unidirectional from the initiator to the peer. There is only one

STAKeySA with the same initiator and peer MAC addresses. The STAKeySA consists of the STAKey, pairwise cipher suite selector, initiator MAC address, and the peer MAC address.

12.4.6 Discovery Process

The most important thing of course is that the STA recognizes the AP and connects to it. This is what we call the discovery process. Each AP advertises its capabilities in the beacon, and probes a response. The detailed process is shown in Figure 12.10. After discovery the STA is ready to perform authentication. Once authentication is done, the keys are generated, after which the port is opened for data transfer.

12.4.7 Preauthentication

In preauthentication the STA can be authenticated with multiple APs at a time. These APs may or may not be in the radio range of the STA. The result of

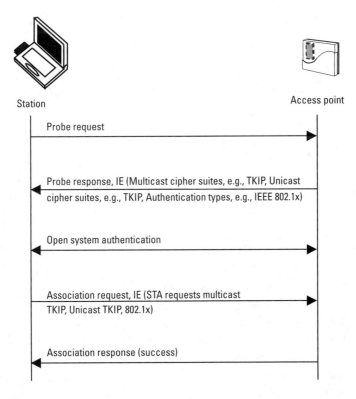

Figure 12.10 Discovery process.

preauthentication may be a PMKSA, if the IEEE 802.1X authentication completes successfully. If preauthentication produces a PMKSA, then, when the supplicant (STA) associates with the preauthenticated AP, the STA can use the PMKSA with the four-way handshake. The PMKSA is inserted into the PMKSA cache. If the STA and AP lose synchronization with respect to the PMKSA, the four-way handshake will fail. Even if a STA has preauthenticated, it is still possible that it may have to undergo a full IEEE 802.1X authentication, as the AP may have purged its PMKSA due to, for example, unavailability of resources or delay in the STA associating.

12.4.8 TKIP

TKIP was developed to provide an intermediary solution until the AES solution was available and to be usable with the existing hardware [1, 4, 29, 30]. Thus TKIP can counter most of the security issues in WEP. The mechanisms that TKIP makes use of are the MIC value called Michael, the extended IV as TSC, and encryption using RC4. TKIP is described in Chapter 3.

12.4.9 CCMP

CCMP provides confidentiality, integrity, data origin authentication and replay protection. Use of protocol is necessary for adequate security. CCMP is described in Chapter 3.

12.4.10 Independent Basic Service Set

IEEE 802.11i security in independent basic service set (IBSS) is discussed briefly next. Note that in IBSS, each STA can act as both a supplicant and authenticator.

1. A shared key, PMK, is decided and distributed among all members. It can be verbal.
2. Start IBSS and use standard IBSS procedure (i.e., the beacon is sent by a STA at a time). The beacon-sending procedure is based on the backoff method. The STA of which the backoff ends first sends the beacon. This continues until the network is alive.
3. PTK is created with a four-way handshake between STAs desiring to communicate. The STA with the smallest MAC becomes the IEEE 802.1X supplicant, while the other STA becomes the authenticator.
4. GTK is created with all the STAs in the IBSS using the two-way handshake.

This procedure does provide security but at the cost of complexity in terms of number of keys to be stored.

12.5 Secure WLAN Deployment

In this section, secure WLAN deployment method for corporate, WISPs and mobile operators are given [32, 33].

12.5.1 Corporate WLAN Deployment

WLAN technology is applicable to all enterprises. Although the IEEE 802.1X EAP is considered the optimal deployment model (see Chapter 10) for most enterprise-particular markets, such as the finance sector, corporate users are likely to prefer triple data encryption standard (3DES) IPsec-based security deployment models. Vertical applications within an organization may be using application-specific clients that can only support static WEP, and these clients will use the static WEP security deployment model. The security model has the largest impact upon the network design. There are three security models presented here:

- IEEE 802.1X EAP security model;

- IPsec VPN security model;

- Static WEP security model.

The security design solutions presented in this section have the following characteristics:

- The security solution model depends upon the security requirements for the corporate LAN. The focus here is on the two most secure solutions, namely, EAP and IPsec VPN.

- Where EAP or IPsec VPNs are not possible, static WEP and access filtering are discussed, although they are not a recommended security deployment model.

- The designs assume that one security model is used (i.e., EAP, IPsec, or Static WEP are not mixed within the one enterprise).

- WLAN APs should be on a dedicated subnet (not shared with wired LAN users).

- The wired LAN is not replaced by the WLAN. The WLAN is used to enhance the current network flexibility and accessibility by providing an extension to the existing network.

- Local area mobility within the corporate LAN is required.

12.5.1.1 IEEE 802.1X EAP Deployment

Such deployment addresses users that are operating in *nomadic* mode; these clients move from place to place and need network connectivity when they are stationary. For example, employees going from their desks to a meeting room normally do not access the network while traveling to the meeting room, but would need access once there. Figure 12.11 shows the EAP implementation of WLAN as dynamic layer 2 key generation with RADIUS. This delivers the features of the ideal WLAN with only the addition of RADIUS authentication

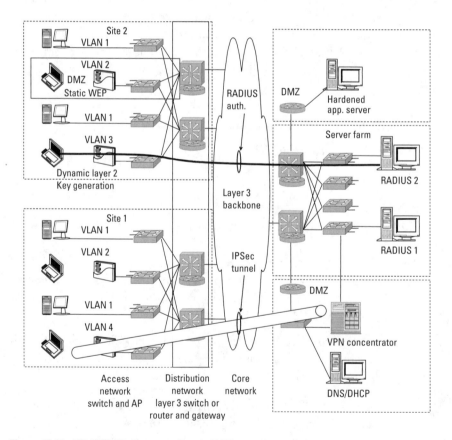

Figure 12.11 WLAN EAP, IPsec, and static WEP security model.

servers (see Chapter 9). The figure is given with VLANs, which provide further security including access control. EAP is considered the optimal solution because of the following reasons:

- Provides per user authentication and accounting;
- Provides dynamic layer 2 key that can be used as WEP key or for IEEE 802.11i;
- No additional filtering or access control required;
- Multiprotocol support and may carry protocols other than IP over the WLAN;
- Filtering requirements at the network access layer are the same as those for wired implementations.

While EAP is the recommended option, it may not be suitable in all cases for the following reasons:

- EAP requires EAP-aware APs and WLAN clients.
- Security features offered by IPsec, such as 3DES encryption, OTP support, or per-user policies, are desired.
- Where seamless roaming within a layer-2 domain is required, EAP clients may take longer to roam between APs; compared to those using static WEP, this may impact solutions such as VoIP over 802.11. (See Chapter 5.)

12.5.1.2 IPsec Deployment

The IPsec solution requires users to connect to the network through a VPN client, even though they are within the campus. A schematic of this is shown in Figure 12.11. Following are the characteristics of a WLAN using IPsec VPNs:

- WLAN with IPsec extension does not require the use of EAP and allows any client adaptor to be used with a 3DES encryption.
- It allows the use of multifactor authentication systems OTP systems.
- It requires the implementation of extensive filters on the network edge to limit network access to IPsec-related traffic destined to the VPN concentrator network.

- It requires user intervention (i.e., the users have to launch the VPN client before they attach to the network).

- Local traffic must still go through the VPN concentrator in the demilitarized zone (DMZ), causing traffic to cross the network multiple times, increasing traffic across the network and degrading performance.

12.5.1.3 Static WEP Deployment

WLAN static WEP addresses the specialized clients that are application-specific that support only static WEP. Within each enterprise, small application verticals exist that can benefit from WLAN applications. Applications requiring this type of solution may also require uninterrupted seamless coverage, as they are special-ized mobility applications. Examples of potential WLAN applications that may use static WEP are

- VoIP over 802.11;

- Messaging applications;

- Workflow and security applications.

Figure 12.11 shows the WLAN static WEP network. The DMZ notation indicates additional filtering required for securing the network, and the inclusion of the layer 2 backbone indicates a possible need to extend the layer 2 network to support campuswide roaming. WLAN static WEP, shown in Figure 12.11, is a design that supports clients who are incapable of EAP or IPsec. The solution is considered less satisfactory than EAP or IPsec for the following reasons:

- Wireless privacy is provided by static WEP, which is vulnerable to attacks; the higher protocol layers should provide additional privacy.

- It introduces logistical problems with key management.

- It requires the implementation of extensive filters on the network edge to limit access to vertical-application-related traffic destined for the clients and servers.

- It may require the use of a firewall to secure the application protocols used.

- It requires that the application server be hardened to prevent attacks from WLAN.

12.5.1.4 Selection Criteria Model

Table 12.1 summarizes the deployment solutions discussed earlier. Table 12.2 shows a detailed summary of the different security deployment options in the WLAN solution space, with regard to privacy and network access.

Table 12.1
Characteristics of WLAN Security Deployment Solutions

	EAP	IPsec	Static WEP
Protocols	Multiprotocol	IP unicast only	Multiprotocol
NIC cards	WPA- or 802.11i-compliant	802.11b-compliant	802.11b-compliant
Connection to network	Integrated with Win login; others enter user id/password	The user must launch a VPN client and login	Transparent to user
Clients	Laptops and high-end PDAs; range of OSs	Laptops and high-end PDAs; range of OSs	Any 802.11 client
Authentication	Username/password	OTP or username/password	Matching WEP key required
Privacy	Dynamic, WEP with time-limited keys and TKIP enhancements	3DES	Problematic key management
Impact on existing network architecture	Additional RADIUS Server required	Additional infrared WLAN will be on a DMZ and require VPN concentrators, authenticated servers, and DHCP servers	Option of additional firewall software or hardware at access layer
Filtering	None required	Extensive filtering required, limiting network access until VPN authentication has occurred	Extensive filtering required, limiting wireless access to only certain predetermined applications
Layer 2 roaming	Transparent—automatically reauthorizes without client intervention (may be slower than VPN or WEP)	Transparent—may be easier to extend layer 2 domain, due to reduced broadcast and multicast traffic	Transparent
Layer 3 roaming	Requires IP address release/renew or MIP solution	Requires IP address release/renew or MIP solution	Requires IP address release/renew or MIP solution

Table 12.1 (continued)

	EAP	IPsec	Static WEP
Management	Network is open to existing network management systems	Filtering must be adjusted to support management applications	May have application-specific management requirements; filtering must be adjusted to support the management applications
Multicast	Supports multicast	Currently cannot support IP multicast	Supports multicast
Performance	WEP encryption performed in hardware	3DES performed in software; throughput will be degraded	WEP encryption performed in hardware

12.5.1.5 Corporate WLAN Deployment Issues

WLAN APs that are found today in the corporate or enterprise environment are individually responsible for traffic handling, radio frequency management, mobility, and to some extent authentication (e.g., static WEP) [3]. These APs act in isolation, making it difficult to perform critical functions such as secure seamless roaming, single sign-on, load balance among other APs, QoS support, and radio management across the entire network. In a small-scale deployment this is fine, but in an enterprise environment where there are tens and hundreds of APs supporting hundreds of users with distributed control function assigned to individual APs, it is a major challenge for network managers to guarantee secure seamless mobility.

In order to guarantee function such as mobility across subnets, QoS, security, traffic policing, and load balancing across wireless and wired networks, it is necessary for the APs to act more like a wireless access server (WAS) working at the networking layer instead of simple layer 2 devices. Such APs (or WAS) should have greater memory and processing power and should be capable of acting as a policy enforcement point in order to guarantee the SLA across the entire corporate network. Further, there is also a great need for a centralized network management system to deal with traffic management, authentication, encryption, and policy decisions (see Figure 12.12).

Although a WAS-based solution is one line of thought, there is another line of thought in the industry that is aiming at a split MAC solution leading to thin APs. The idea is to move several core functionalities to a switch from where the control and management can be done. This switch is known as the access controller (AC). Such switches will also serve as policy enforcement points

Table 12.2
Encryption and Network Access Options

	EAP (EAP-TLS)	IPsec	Static WEP
Key length (bits)	128	168	128
Encryption algorithm	RC4	3 DES	RC4
Packet integrity	CRC32/MIC	MD5-HMAC/SHA-HMAC	CRC32/MIC
Device authentication	None	Preshared secret or certificates	None
User authentication	Username/password (PKI certificates)	Username/password or OTP	None
User differentiation	No	Yes	No
Transparent user experience	Yes	No	Yes
ACL requirements	None	Substantial	N/A
Additional hardware	Authentication server (certificate authority)	Authentication server and VPN gateway	No
Per users keying	Yes	Yes	No
Protocol support	Any	IP unicast	Any
Client support	PCs and high-end PDAs	PCs and high-end PDAs	All clients supported
Open standard	No	Yes	Yes
Time-based key rotation	Configurable	Configurable	No
Client hardware encryption	Yes	Available, software is most common method	Yes
Additional software	No	IPsec client	No
Per-flow quality of service (QoS) policy management	At access switch	After VPN gateway	At access switch

(PEPs). A study going on in the IETF control and provisioning of wireless access point (CAPWAP) group gives several possible WLAN network architectures. The goal of CAPWAP is to produce solutions for centralized management with intelligence at the switch instead of at the AP. AP is termed a wireless termination point (WTP). The schematic for a CAPWAP type network will be similar to Figure 12.12 except that the WAS should be WTP and thus without the router. The switches to the APs (WTPs) will be the ACs and can also be working in layer 3.

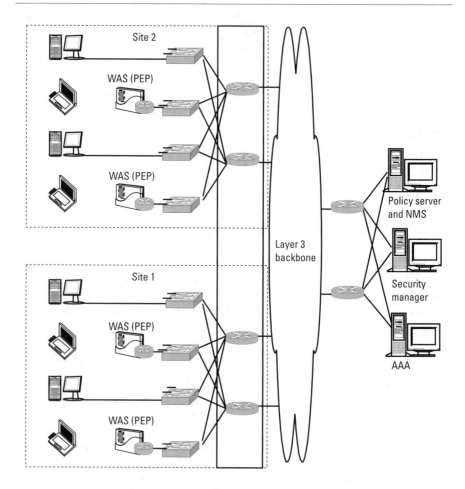

Figure 12.12   Corporate WLAN security using centralized security manager, AAA, and policy server.

12.5.2 Public WLAN Deployment

There are two types of public WLAN (PWLAN) hotspots, secure and nonsecure networks. Nonsecure PWLAN is based on the notation of just providing Internet access to the customers, and if a corporate user requires secure network access to his/her corporate network then the user needs to initiate an IPsec tunnel from his/her device to the corporate VPN gateway. An alternative method is to run a Web-based application layer SSL VPN. The latter is the most cost-effective solution, and it suits an organization that mainly uses Web applications for its business. The other advantage of SSL VPN is that it does not require client software for the end user device, nor does it require negotiating SA in advance with the corresponding entity.

Deployment of hundreds of secure, cost-efficient PWLAN hotspots requires a service provider to take countermeasures against the security threat by WLAN and exposure of user data and network entities to potential hackers. Figure 12.13 shows a centralized security solution for a PWLAN hotspot using digital subscriber line (DSL) as backhaul transmission. A centralized architecture is chosen over a decentralized because remote network management, such as a change in security policy or a software upgrade from the central service area, is faster and more convenient than going out to change settings at every hotspot location. In the centralized security architecture, the link between the hotspot AR and management router at the central service area is secured using IPsec ESP 3DES encryption [9]. This secure link is dedicated for network management and for operational support system. The secure link between the hotspot AR and edge access router (EAR) located at the central service area is secure using IPsec ESP null instead of AH. Since IPsec ESP null provides integrity protection and authentication services without covering the IP header in the message integrity check, IPsec ESP null is used in preference to IPsec AH. This link secures the traffic between the hotspot AR and the service provider's centralized service area. However, the link between the AP and end-user device is exposed to a potential security threat from hackers, since WEP encryption is easy to crack.

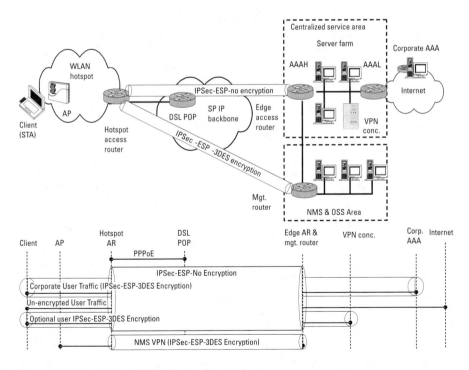

Figure 12.13 Centralized PWLAN security solution.

For a corporate user, this is no issue because an end-to-end secure tunnel can be set between the client device and corporate VPN gateway. All other users may gain secure access between the client device and the central service area VPN gateway by creating an IPsec ESP 3DES encryption tunnel within a secure IPsec ESP null tunnel existing between AR and EAR (shown in Figure 12.13) for secure link between the client device and service provider's central service area. For this scenario the client is required to install preconfigured IPsec client software, which should be made available by the wireless service provider.

12.5.3 Operator-Owned PWLAN Solutions

For the deployment of PWLAN by mobile operators, there are two fundamental interworking solutions, tight and loose interworking, and it depends on the level of integration required between the systems (see Figure 12.14). The tight interworking solution is based on the idea of making use of the WLAN radio interface as a bearer for a cellular network (e.g., GPRS/enhanced data for GSM evolution (EDGE) /UMTS), with all the network control entities in the core network integrated. A tight interworking solution would mandate the full 3GPP security architecture and require the 3GPP protocol stacks and interfaces to be present in the WLAN system. For a loose interworking solution, there is no need to make changes to the WLAN standard. This solution has the benefit of not needing a convergence layer and avoids link layer modifications; the authentication protocol is allowed to run at the link layer using EAP and AAA as transport mechanism. A fundamental requirement in 3GPP is that 3GPP-WLAN interworking shall not compromise the UMTS security architecture. Therefore, it is required that the authentication and key distribution be based on the EAP-enhanced GSM authentication (EAP-SIM) or EAP UMTS authentication and key agreement (EAP-AKA) (see Chapter 15).

SIM-based authentication (EAP-SIM/802.1X) is based on open standards and allows a GSM operator to leverage its existing subscriber security database and network elements. The approach is fully aligned with the RSNs being defined by the IEEE 802.11i. This includes the reuse of the GSM A8 algorithm for a secure per-user, per-session key exchange for implementing encryption over the WLAN air interface. This standard allows users to roam between different 802.1X WLAN networks, as the specifics of the SIM-based security mechanism are transparent to the visited WLAN network. Finally, SIM authentication based on EAP-SIM/802.1X does not require new WLAN network interface cards (NICs). Instead, the installed base of WLAN enabled 802.1X PCs and PDAs can be SIM-enabled with the simple addition of a SIM reader using the industry-standard PC/smart card interface or connection between GSM handset and mobile client through infrared, cable, or Bluetooth. This discussion is also valid for EAP-AKA.

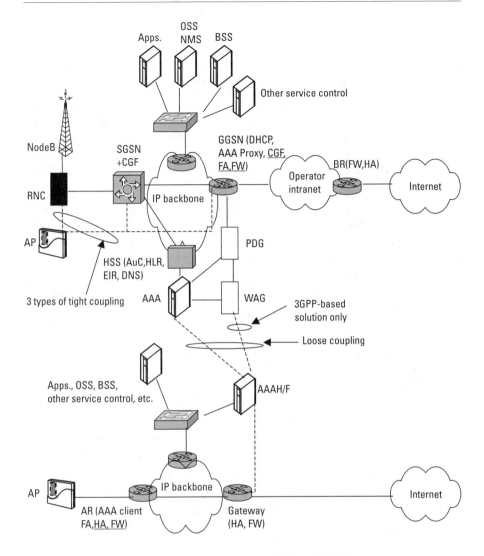

Figure 12.14 No, loose, and tight coupling between 3GPP and WLAN networks.

Another possibility for the operator is to use SMS. Both SMS-based and SIM-based PWLAN deployment by an operator is discussed next.

12.5.3.1 SMS-Based Public WLAN Deployment

Mobile operators may offer PWLAN access to postpaid clients on a monthly subscription for limited or unlimited usage, depending on the type of subscription. In order to offer granular WLAN access to prepaid, postpaid, and roaming users, operators may use SMS bearer to communicate OTP to its customer. (See

Figure 12.15.) As for the prepaid users, it is important to check their credit balances before WLAN access is granted. Customers could also pay by credit card if they are connected to a foreign operator with no roaming agreement; in this case a Web server could be used. The Web server should communicate in some form or other with the credit card company to debit the amount. Once the amount is debited successfully, the user can access the PWLAN. A customer of mobile operators where a roaming agreement exists sends a SMS with the request of WLAN usage for a given time period. The OTP server receives the request and contacts the billing/prepaid charging system to check if the customer has an adequate deposit on his account. If the customer has adequate deposit, the OTP server computes a OTP and sends it via SMS to the customer; if not, an error message will be returned. Afterward, the OTP server sends the password to an AAA server via LDAP. If the customer uses the password, the AAA server tells the OTP server the use of the password, and the OTP server tells the billing system the amount a user has to pay. Figure 12.15 shows the concept of using short message service center (SMSC) and OTP server to authenticate, authorize, and bill users for WLAN usage. Figure 12.16 illustrates the procedure of a successful transaction of a prepaid user for its WLAN usage.

12.5.3.2 SIM-Based Public WLAN Deployment

As service providers start to deploy PWLAN, they are finding well-known issues with techniques that do not incorporate IEEE 802.11 encryption. In particular,

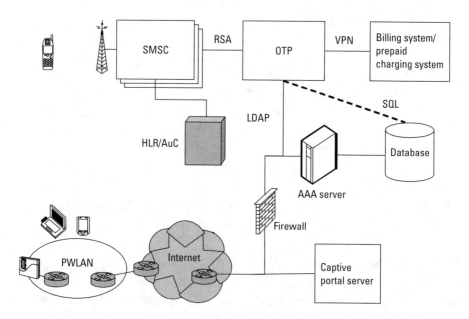

Figure 12.15 SMS-based PWLAN deployment.

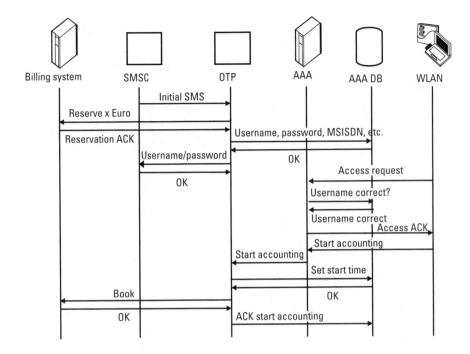

Figure 12.16 SMS-based prepaid procedure of a successful transaction.

session hijacking cannot be prevented if encryption is not enabled or supplementary security is applied (e.g., a client-initiated IPsec tunnel). This is becoming an issue with service provider deployments, as they realize that they risk a tarnished image due to any bad press generated due to such well-known weaknesses. SIM-based security deployment is interesting for a GSM operator for the following reasons:

- It leverages an already deployed SA using a high-entropy shared key, *Ki*, stored in a tamper-proof smart card.
- It leverages already deployed authentication and cipher key generation algorithms, A3 and A8.

Regarding new security threats compared with GSM, the rogue network is not considered a security threat for GSM but is clearly a threat with WLAN. In order to combat this, EAP-SIM enhances GSM A3 A8 security with network authentication. With regard to securing the WLAN 802.11 link, EAP-SIM reuses the A8 algorithm to support mutual key exchange for the negotiated cipher suite. According to the negotiated cipher suite, additional protection may be required to guard against initialization vector rollover. Hence, rekeying should be performed before rollover. In such cases, the AAA session timeout attribute in the access accept is set to ensure rekeying before rollover.

12.5.3.3 Mobile and WLAN Roaming

Delivering a roaming service encompasses a number of key building blocks: signaling, billing, data clearing, financial clearing and settlement, contract management, testing, and fraud management. Because of the significant investment made by the mobile service providers in transferred account procedure (TAP), it is clearly advantageous to look to leverage this core competency when looking to build a WLAN roaming service. While much of the interest has been to use SIM-based authentication for WLAN users in order to enable roaming, this is not strictly necessary. It is highly unlikely that a single authentication standard will exist for all roaming users. Therefore, a generic mapping from international mobile subscriber ID (IMSI) to non-IMSI is required even for EAP-SIM authentication; this would allow an existing GPRS ticket for WLAN users. This then can allow an operator other than the home network to perform mediation of AAA-based accounting records into TAP tickets. These two techniques, when combined, offer the capability of using any authentication mechanism, including user name and password, the use of which can be mediated into a GPRS ticket. The choice of which authentication mechanisms to be supported and the mapping between non-IMSI and IMSI is then a matter for the home operator.

AAA (e.g., RADIUS and Diameter) protocols are peer to peer. Both require shared secrets and both require secure transport. Both 3GPP and WLAN interworking using AAA architecture, as shown in Figure 12.14. Hence, there are recognized scalability issues with roaming using such functionality (e.g., compared with over 30,000 existing GSM bilateral roaming agreements). Consequently, scalable AAA requires broker functionality. Such a broker will have bilateral peering agreements with a number of WLAN providers and also with home networks offering roaming WLAN service to their subscribers. This roaming broker functionality can be performed by an established GSM operator, a GRX provider, a GSM clearinghouse, or a new entity.

One of the clear advantages of roaming broker functionality is that it allows the AAA interfaces to scale. One of the disadvantages is that it means that the home operator no longer has a direct (business) relationship with the WLAN provider. This is important because the WLAN provider may not be trusted by the home network provider. Hence, the broker must assume the risk of validating the correct operation of the interoperator interfaces between the broker and the WLAN provider. This may include validating that the accounting records produced by the WLAN provider are authentic. Depending on both the interoperator billing metric (time, volume, or flat fee) and the level of trust between the broker and the WLAN provider, this assumption of risk by the roaming broker may mean that broker equipment is placed in the user plane for roaming subscribers supported using this broker. This equipment can then be used to detect fraud by the WLAN provider.

12.5.4 Secure Network Management

SNMP [34–37] provides management capabilities for transport control protocol/Internet protocol (TCP/IP)–based networks, and because of its simplicity and the achievement of interoperability of the SNMP module from different vendors it became a de facto standard for the management of network-based equipment. SNMPv3 [37] includes three important services: authentication, privacy, and access control, as shown in Figure 12.17. To deliver these services in a flexible and efficient manner, SNMPv3 introduces the concept of a principal, which is the entity on whose behalf services are provided or processing takes place. A principal can be an individual acting in a particular role; a set of individuals each acting in a particular role; an application or set of applications; or a combination thereof. In essence, a principal operates from a management station and issues SNMP commands to agent systems. The identity of the principal and the target agent together determine the security features that will be invoked, including authentication, privacy, and access control. The use of

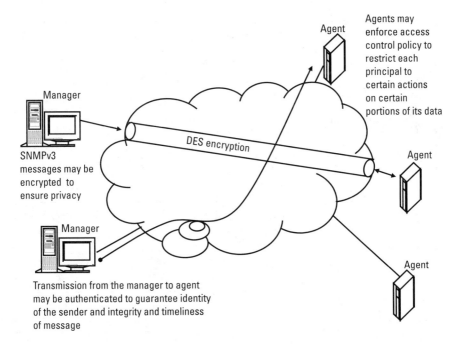

Figure 12.17 SNMPv3 security features.

principals allows security policies to be tailored to the specific principal, agent, and information exchange, and gives human security managers considerable flexibility in assigning network authorization to users.

SNMPv3 is defined in a modular fashion, as shown in Figure 12.18. Each SNMP entity includes a single SNMP engine. An SNMP engine implements functions for sending and receiving messages, authenticating and encrypting/decrypting messages, and controlling access to managed objects. These functions are provided as services to one or more applications that are configured with the SNMP engine to form an SNMP entity. This modular architecture provides several advantages. First, the role of an SNMP entity (see Table 12.3) is determined by the modules that are implemented in that entity. For example, a certain set of modules is required for an SNMP agent, whereas a different (though overlapping) set of modules is required for an SNMP manager. Second, the modular structure of the specification lends itself to defining different versions of each module. This, in turn, makes it possible to (1) define alternative or enhanced capabilities for certain aspects of SNMP without needing to go to a new version of the entire standard (e.g., SNMPv4), and (2) clearly specify coexistence and transition strategies.

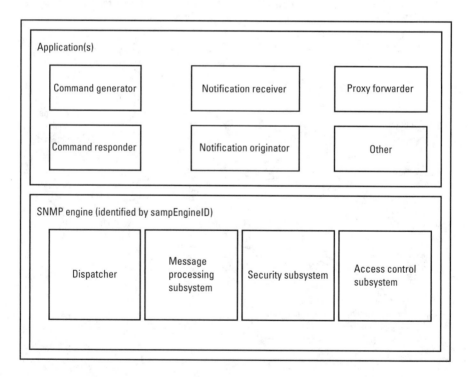

Figure 12.18 SNMP entity (RFC 2271).

Table 12.3
Components of SNMP Entity (RFC 2273)

Component	Description
Dispatcher	Allows for concurrent support of multiple versions of SNMP messages in the SNMP engine. It is responsible for (1) accepting protocol data units (PDUs) from applications for transmission over the network and delivering incoming PDUs to applications; (2) passing outgoing PDUs to the message processing subsystem to prepare as messages, and passing incoming messages to the message processing subsystem to extract the incoming PDUs; and (3) sending and receiving SNMP messages over the network.
Message processing subsystem	Responsible for preparing messages for sending and for extracting data from received messages.
Command responder	Receives SNMP Get, GetNext, GetBulk, or Set request PDUs destined for the local system as indicated by the fact that the contextEngineID in the received request is equal to that of the local engine through which the request was received. The command responder application performs the appropriate protocol operation, using access control, and generates a response message to be sent to the originator of the request.
Security subsystem	Provides security services such as the authentication and privacy of messages. This subsystem potentially contains multiple security models.
Access control subsystem	Provides a set of authorization services that an application can use for checking access rights. Access control can be invoked for retrieval or modification request operations and for notification generation operations.
Command generator	Initiates SNMP Get, GetNext, GetBulk, or Set request PDUs and processes the response to a request that it has generated.
Notification originator	Monitors a system for particular events or conditions and generates trap or inform messages based on these events or conditions. A notification originator must have a mechanism for determining where to send messages, and which SNMP version and security parameters to use when sending messages.
Notification receiver	Listens for notification messages and generates response messages when a message containing an inform PDU is received.
Proxy forwarder	Forwards SNMP messages. Implementation of a proxy forwarder application is optional.

12.5.4.1 Secret Key Authentication

The authentication mechanism in SNMPv3 assures that a received message was, in fact, transmitted by the principal whose identifier appears as the source in the message header (see Figure 12.19). In addition, this mechanism assures that the message was not altered in transit and that it was not artificially delayed or replayed. To achieve authentication, each pair of principal and remote SNMP engines that wishes to communicate must share a secret authentication key. The sending entity provides authentication by including a message authentication code with the SNMPv3 message it is sending. This code is a function of the contents of the message, the identity of the principal and engine, the time of transmission, and a secret key that should be known only to the sender and the receiver. The secret key must initially be set up outside of SNMPv3 as a configuration function. That is, the configuration manager or network manager is

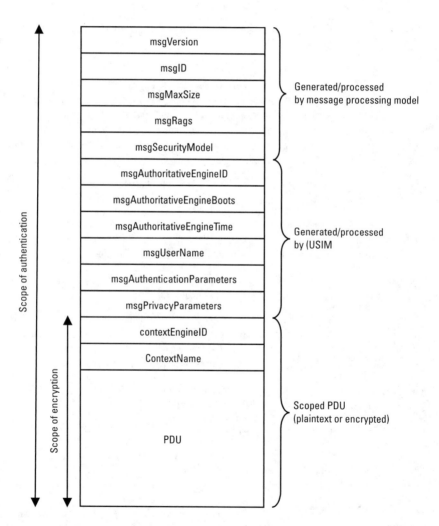

Figure 12.19　SNMPv3 message format.

responsible for distributing initial secret keys to be loaded into the databases of the various SNMP managers and agents. This can be done manually or by using some form of secure data transfer outside of SNMPv3. When the receiving entity gets the message, it uses the same secret key to calculate the message authentication code again. If the receiver's version of the code matches the value appended to the incoming message, then the receiver knows that the message can only have originated from the authorized manager and that the message was not altered in transit. The shared secret key between sending and receiving parties must be preconfigured.

User-based security model (USM) authentication is used for timeliness verification. USM is responsible for assuring that messages arrive within a reasonable time window to protect against message delay and replay attacks. Two functions support this service: synchronization and time-window checking. Each authoritative engine maintains two values, snmpEngineBoots and snmpEngineTime, that keep track of the number of boots since initialization and the number of seconds since the last boot. These values are placed in outgoing messages in the fields msgAuthoritativeEngineBoots and msgAuthoritativeEngineTime. A nonauthoritative engine maintains synchronization with an authoritative engine by maintaining local copies of snmpEngineBoots and snmpEngineTime for each remote authoritative engine with which it communicates. These values are updated on receipt of an authentic message from the remote authoritative engine. Between these message updates, the nonauthoritative engine increments the value of snmpEngineTime for the remote authoritative engine to maintain loose synchronization. These values are inserted in outgoing messages intended for that authoritative engine. When an authoritative engine receives a message, it compares the incoming boot and time values with its own boot and time values. If the boot values match and if the incoming time value is within 150 seconds of the actual time value, then the message is declared to be within the time window and, therefore, to be a timely message.

12.5.4.2 Privacy Using Conventional Encryption

The SNMPv3 USM privacy facility enables managers and agents to encrypt messages to prevent eavesdropping by third parties. Again, manager entity and agent entity must share a secret key. When privacy is invoked between a principal and a remote engine, all traffic between them is encrypted using the DES. The sending entity encrypts the entire message using the DES algorithm and its secret key and sends the message to the receiving entity, which decrypts it using the DES algorithm and the same secret key. Again, the two parties must be configured with the shared key. The cipher-block-chaining (CBC) mode of DES is used by USM. This mode requires that an initial value be used to start the encryption process. The msgPrivacyParameters field in the message header contains a value from which the initial value can be derived by both sender and receiver.

12.6 802.11i Future

The IEEE 802.11i MAC security enhancements provide sufficient means to secure WLAN networks from multiple attacks. The standard was approved in April 2004, but still a lot of deployed devices are legacy devices and do not support the latest security enhancements. A standard is considered secure after undergoing public scrutiny. As more 802.11i-compliant hardware are installed, attackers will attempt to break the system.

There are attacks that the IEEE 802.11i specification does not address. The major one among these is DoS due to nonprotection of management packets. As association frames are not encrypted and integrity protected, anyone could send a disassociation message and choose the sender's address, thus causing the station to be cut off from communication with the AP. Some security issues also remain when legacy devices are used with a software patch to support TKIP instead of CCMP, but this is a calculated risk as it is known that TKIP only provides a temporary solution before all hardware can be replaced.

One of the major problems in security is that, even when available, users disable security features for ease of use. All security options in the 802.11i standard are optional, and the default configuration is no security. Although WEP is broken and multiple applications are available on the net to break a WEP-protected system, users are advised to enable it to discourage attackers that are simply looking for nonprotected access and who wouldn't attempt to breach a system with a minimum of protection. Whether WEP, TKIP, or CCMP is used, key distribution must be performed before encryption can be initiated. Automatic key distribution will probably be set up in corporate organizations, whereas private users will likely still depend on manual distribution, which is considered a burden by many. Last but not least, enabling security features affects performance, as it takes time and computational power to execute cryptographic algorithms. While developers are doing their best to optimize performance and reduce the charge on users, a certain effort cannot be avoided. Users must be aware of the compromise they are making by choosing ease of use and performance versus security.

References

[1] ISO/IEC 8802-11, ANSI/IEEE Std 802.11, "First Edition 1999-00-00, Information Technology—Telecommunications and Information Exchange Between Systems—Local and Metropolitan Area Networks—Specific Requirements—Part 11: Wireless LAN Medium Access Control (MAC) and Physical Layer (PHY) Specifications."

[2] Prasad, A. R., "WLANs: Protocols, Security and Deployment," Ph.D. Thesis, Delft, the Netherlands: Delft University Press, December 2003.

[3] Prasad, N. R., and A. R. Prasad, (eds.), *WLAN Systems and Wireless IP for Next Generation Communications,* Norwood, MA: Artech House, 2002.

[4] Edney, J., and W. A. Arbaugh, *Real 802.11 Security: Wi-Fi Protected Access and 802.11i,* Reading, MA: Addison-Wesley, 2004.

[5] Arbaugh, W.A., http://www.cs.umd.edu/~waa/wireless.html.

[6] Borisov, N., I. Goldberg, and D. Wagner, "Intercepting Mobile Communications: The Insecurity of 802.11," *7th Annual International Conference on Mobile Computing and Networking,* July 2001, Rome, Italy.

[7] Walker, J. R., "Unsafe at Any Key Size: An Analysis of the WEP Encapsulation." IEEE 802.11/00-362, October 2000.

[8] Fluhrer, S., I. Martin, and A. Shamir, "Weaknesses in the Key Scheduling Algorithm of RC4," *8th Annual Workshop on Selected Areas in Cryptography,* August 2001, Toronto, Canada.

[9] Stubblefield, A., J. Ioannidis, and A. D. Rubin, "Using the Fluhrer, Mantin and Shamir Attack to Break WEP," *Network and Distributed Systems Security Symposium 02,* 2002, San Diego, CA.

[10] Burrows, M., M. Abadi, and R. Needham, "A Logic of Authentication," *ACM Transactions on Computer Systems,* Vol. 8, No. 1, February 1990.

[11] Arbaugh, W. A., N. Shankar, and J. Wang, "Your 802.11 Network Has No Clothes," *Proc. 1st IEEE International Conference on Wireless LANs and Home Networks,* Singapore, December 2001, pp. 131–144.

[12] Cam-Winget, N., et al., "Security Flaws in 802.11 Data Link Protocols," *Communications of the ACM,* Vol. 46, No. 5, May 2003, pp. 34–39.

[13] Housley, R., and W. A. Arbaugh, "Security Problems in 802.11-Based Networks," *Communications of the ACM,* Vol. 46, No. 5, May 2003, pp. 31–34.

[14] Arbaugh, W. A., "An Inductive Chosen Plaintext Attack against WEP/WEP2," IEEE Document 802.11-02/230, May 2001.

[15] Petroni, Jr., N. L., and W. A. Arbaugh, "The Dangers of Mitigating Security Design Flaws: A Wireless Case Study," *IEEE Security and Privacy Magazine,* Vol. 1, No. 1, January–February 2003, pp. 28–36.

[16] UNINETT, http://www.uninett.no/wlan/wlanthreat.html#02, June 2, 2004.

[17] NetStumbler, http://www.netstumbler.com/download.php?op=viewdownload&cid=1& orderby=hitsD.

[18] Kismet, http://www.kismetwireless.net.

[19] Ethereal, http://www.ethereal.com/http://www.uninett.no/wlan/wlanthreat.html#02.

[20] SMAC, http://www.klcconsulting.net/smac.

[21] GNU MAC Changer, http://www.alobbs.com/modules.php?op=modload&name=macc& file=index.

[22] Airsnort, http://airsnort.shmoo.com.

[23] WEPCrack, http://wepcrack.sourceforge.net.

[24] AirJack, http://www.11ninja.net.

[25] Danielyan, E., "The Lure of Biometrics," *Cisco IP Journal,* Vol. 7, No. 1, March 2004, pp. 15–34.

[26] VPNC, http://www.vpnc.org.

[27] IDS, http://www.intrusion-detection-system-group.co.uk.

[28] Prasad, A., and A. Raji, "A Proposal for IEEE 802.11e Security," IEEE 802.11e, 00/178, July 2000.

[29] Wi-Fi Protected Access, http://www.wi-fi.org/opensection/protected_access.asp.

[30] Whiting, D., R. Housley, and N. Ferguson, "Counter with CBC-MAC (CCM)," RFC 3610, September 2003.

[31] IEEE 802.11i, "IEEE 802.11: Specification for Robust Security, D10.0," April 2004.

[32] Prasad, N. R., "Adaptive Security for Heterogeneous Networks," Ph.D. Thesis, University of Rome, Rome, Italy, 2004.

[33] Prasad A. R. and N. R. Prasad, *802.11 WLANs and IP Networking: Security, QoS, and Mobility*, Norwood MA: Artech House, 2005.

[34] Wijnen, B., D. Harrington, and R. Presuhn, "An Architecture for Describing SNMP Management Frameworks," RFC 2571, April 1999.

[35] Case, J., et al., "Message Processing and Dispatching for the Simple Network Management Protocol (SNMP)," RFC 3412, December 2002.

[36] Levi, D., P. Meyer, and B. Stewart, "Simple Network Management Protocol (SNMP) Applications," RFC 3413, December 2002.

[37] Blumenthal, U., and B. Wijnen, "User-Based Security Model (USM) for Version 3 of the Simple Network Management Protocol (SNMPv3)," RFC 3414, December 2002.

13

WMAN

WMAN, particularly based on IEEE 802.16 standard, is gaining grounds. To support multivendor interoperability in 802.16, an industry alliance known as worldwide interoperability microwave access (WiMAX) has been formed. In this chapter, the original 802.16 standard [1–6] is first discussed together with its security issues [2, 3]. Enhancement to the security solutions in the form of 802.16e is discussed next.

13.1 An Introduction to 802.16

A brief introduction to the IEEE 802.16 standard makes sense before discussing the security details. In this section, the 802.16 network design, protocols, and security requirements are presented.

13.1.1 Standard Family

The original IEEE 802.16 standard was approved in December 2001. This standard developed a solution for point-to-multipoint WMAN solutions. As for all 802 standards, the standard defines MAC and physical layers (PHY). This standard requires line of sight (LOS) between base station (BS) and a subscriber station (SS) and works at frequency band of 10–66 GHz. It defines bit rates from 32 to 134 Mbps. IEEE 802.16a was approved in 2003 and enables operation at frequencies of 2–11 GHz. IEEE 802.16a introduces mesh mode and non-LOS (NLOS) service with data rates up to 100 Mbps. IEEE 802.16-2004 consolidates all previous 802.16 standards, while retaining all modes and major features without any additional modules. IEEE 802.16-2004 enhances performance, eases deployment, and corrects the standard and its text. The IEEE 802.16e standard, ratified last year, adds mobility support to 802.16. IEEE 802.16e is also known as wireless broadband (WiBro) in South Korea, and mobile WiMax. WiBro is planned to be deployed in South Korea in 2006, and mobile WiMax in Japan by 2007.

In this chapter, security as defined in IEEE 802.16-2004, and its issues and solutions in draft IEEE 802.16e standard are discussed.

13.1.2   What Is IEEE 802.16?

As stated previously, 802.16 is a WMAN standard gaining ground in the wireless market. Initially conceived as a radio standard to enable cost-effective last-mile broadband connectivity to those not served by wired broadband such as cable or DSL, the specifications are evolving to target a broader market opportunity for mobile, high-speed broadband applications. The promise of realizing a low-cost, broadly interoperable wide area data network that supports portable and mobile usage could have significant end-user benefits. Notably, this network can complement and extend the WiFi hotspot usage model to provide broader IP data service coverage and roaming, which has so far eluded current 3G systems due to system cost and complexity.

The 802.16-2004 [1] standard supersedes all previous versions as the base standard and specifies networks for the current fixed access market segment. The 802.16e [4] amendment and 802.16f and 802.16g task groups will amend the base specification to enable not just fixed, but also portable and mobile operation in frequency bands below 6 GHz.

IEEE 802.16 is optimized to deliver high, bursty data rates to subscriber stations, but the sophisticated MAC architecture can simultaneously support real-time multimedia and isochronous applications such as voice over IP (VoIP) as well. This means that IEEE 802.16 is uniquely positioned to extend broadband wireless beyond the limits of today's Wi-Fi systems, both in distance and in the ability to support applications requiring advanced QoS such as VoIP, streaming video, and online gaming.

The technology is expected to be adopted by different incumbent operator types—for example, WISPs, code division multiple access (CDMA) and wideband CDMA (WCDMA) cellular operators, and wireline broadband providers. Each of these operators will approach the market with different business models, each based on their current markets and perceived opportunities for broadband wireless as well as different requirements for integration with existing (legacy) networks. As a result, 802.16 network deployments face the challenging task of needing to adapt to different network architectures while still supporting standardized components and interfaces for multivendor interoperability.

13.1.2.1 Architecture

Figure 13.1 conceptually depicts the architecture evolution for 802.16. A basic 802.16-2004 based fixed access (indoor[1] and outdoor) deployment is typically accomplished via a static provisioning relationship between a SS and an 802.16 AP. The collection of APs and interconnecting routers or switches comprising the Radio Access Network (RAN) can be logically viewed as a contiguous cloud with no inter-AP mobility requirements from an SS perspective. The RAN(s) interconnect via a logically centralized operator IP core network to one or more external networks as shown. The operator IP core may host services such as IP address management, domain name service, media switching between IP packet-switched data and public switched telephone network (PSTN) circuit-switched data, 2.5G/3G/Wi-Fi harmonization and interworking, and VPN services (provider hosted or transit).

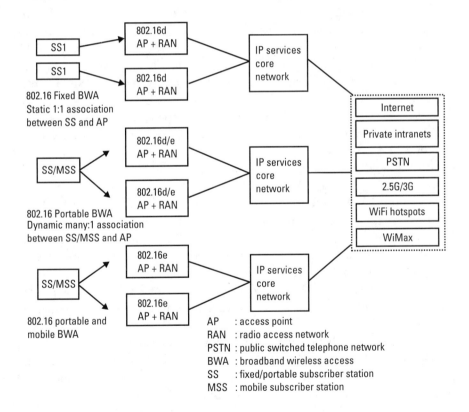

Figure 13.1 802.16 Architecture evolution.

---

1. Indoor operation may require use of beam forming or multiple input multiple output (MIMO) advanced antenna systems (AAS), which is supported in the 802.16 standard.

Going from fixed access to portability with simple mobility involving use of mobile SS (MSS) such as laptops and PDAs introduces network infrastructure changes, such as the need to support break-before-make micromobility and macromobility[2] handovers across APs with relaxed handover packet loss and latency[3] (less than 2 seconds), cross-operator roaming, and the need to support reuse of user and MSS credentials across logically partitioned RAN clouds.

Going from portability to full mobility requires support in the RAN for low (~0) packet loss and low latency (< 100 ms) make-before-break handovers and mechanisms, such as idle mode with paging for extended low-power operation.

An important design consideration is QoS. Fixed access and portable usage models need only support acceptable QoS guarantees for stationary usage scenarios. Portability introduces the requirement to transfer the SLA across APs involved in a handover, although QoS may be relaxed during handovers. Full mobility requires consistent QoS in all operating modes, including handovers. The 802.16 RAN will need to deliver bandwidth and/or QoS on demand as needed to support diverse real-time and nonreal-time services over the 802.16 RAN. Besides the traditional best effort forwarding, the RAN will need to handle latency-intolerant traffic generated by applications such as VoIP and interactive games.

The decoupling of the RAN from an operator IP core network permits incremental migration to fully mobile operation. However, an operator must give due consideration to the RAN topology (such as coverage overlap, user capacity, and range) to ensure that the physical network is future-proof for such an evolution.

Figure 13.2 depicts an end-to-end reference architecture for 802.16. Various functional entities and interoperability interfaces are identified. The network essentially decomposes into three major functional aggregations: the 802.16 SS/MSS, the 802.16 RAN, and interconnect to various operator IP core and application provider networks. The IP core network both manages the resources of the 802.16 RAN and provides core network services such as address management, authentication, service authorization, and provisioning for 802.16 SS/MSSs.

The reference architecture, especially interconnectivity in the RAN and interconnections to remote IP networks, is based on extensive use of native IP protocols that in turn can deliver desired economies of scale. In the following sections, we describe three logical entities: the radio network serving node

---

2. Micromobility refers to handovers between APs within the same IP prefix or subnet domain. Macromobility refers to handovers across APs in different IP prefix or subnet domains.

3. Latency may be unacceptable for real-time IP services such as VoIP during handovers but acceptable for TCP and VPN services as well as store-and-forward multimedia services.

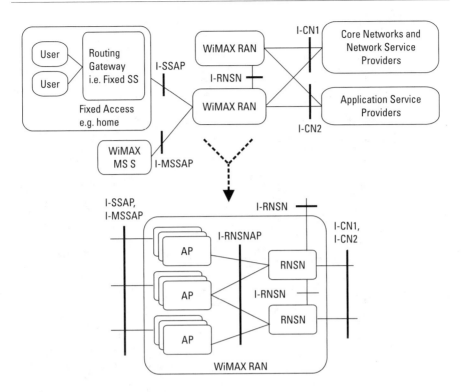

Figure 13.2 802.16 Reference architecture.

(RNSN), AP, and SS/MSS. We also briefly describe the interoperability interfaces identified in Figure 13.2. AP/RAN topologies are depicted in Figure 13.3.

13.1.2.2 Protocol Layers

The protocol layer consists of MAC and PHY layers. The MAC is divided in sublayers: the service-specific convergence sublayer (CS), the MAC common part sublayer (CPS), and the MAC security sublayer. This is shown in Figure 13.4.

The MAC CS basically provides transformation or mapping of data received from the CS service access point (SAP) into MAC service data units (SDUs) that are received by the MAC CPS. The information conveyed includes a MAC service flow identifier (SFID) and connection identifier (CID) and may includea payload suppression header (PSH) function. The PHY consists of various techniques working at different frequency bands.

The MAC CPS provides functions like system access, bandwidth allocation, connection establishment, and connection maintenance. The security sublayer provides authentication, key exchange and encryption. Data, PHY control, and statistics are transferred between MAC CPS and the PHY via the PHY SAP (which is implementation specific).

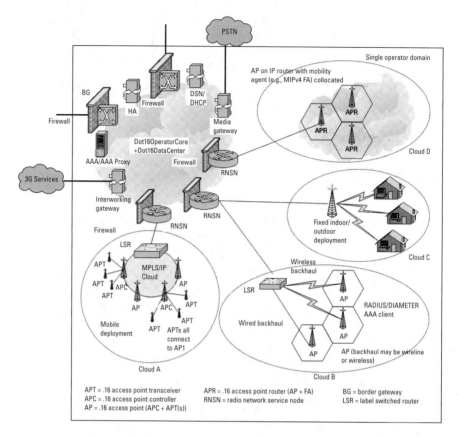

Figure 13.3 802.16 RAN topologies.

13.1.2.3 The MAC Security Sublayer

In 802.16 terms, privacy [4] or security sublayer [1] provide authentication or confidentiality. The sublayer also provides methods to prevent theft of service and unauthorized access. There is also a method for authenticated client/server key management. Use of digital certificates are also provisioned by the standard. The security sublayer constitutes of two component protocols [1–6]:

- *Encapsulation protocol:* This protocol defines the set of supported cryptographic suites; this includes information regarding the pairings of *data encryption* and *authentication* algorithms, and the rules for applying the algorithms to a MAC PDU payload.

- *Key management protocol:* This protocol provides *distribution of keying material* from a BS to SS. The protocol used is the privacy key management (PKM) protocol already deployed in the data over cable service

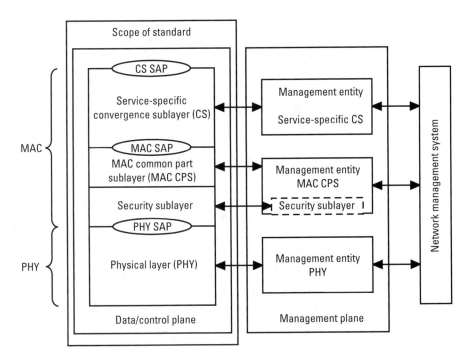

Figure 13.4 IEEE 802.16 protocol layers.

interface specification (DOCSIS)–compliant cable modems [3]. The BS also uses this protocol to *enforce conditional access to the network services.*

The protocol stack is shown in Figure 13.5 [4].

13.2 IEEE 802.16-2004 Security

IEEE 802.16-2004 has the same security sublayer as discussed in Section 13.1.2.3. In this section, the SA and PKM protocols are discussed.

13.2.1 Security Associations

A SA is set between the BS and SS by means of security parameters shared between the two. SAs include encryption keys and initialization vector values. Each SA in 802.16 is identified by a security association identifier (SAID). A BS must ensure that a client SS has access to only the SA which that client SS is authorized to access. Three different types of SAs are defined in the IEEE 802.16-2004 standard [1]:

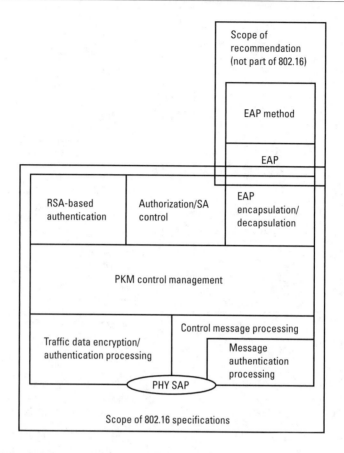

Figure 13.5 IEEE 802.16e MAC security sublayer.

- *Primary SA:* Each SS established a unique primary SA with its BS and a SAID that is equivalent of the basic CID of that SS. The primary SA is established during the SS initialization process.
- *Static SA:* The static SA is established within the BS for the internal purposes of BSs.
- *Dynamic SA:* A dynamic SA is initiated and eliminated as needed in response to the initiation/termination of specific service flows.

A SS requests the keying material from the BS using PKM; it is the responsibility of the BS to give the keys of the SA to which the SS is authorized. The keying material related to a given SA is also assigned a lifetime by the BS, and a given SS is expected to request new keying material from its BS before the current keying material expires using PKM.

In 802.16, for a given SS, all the upstream traffic from the SS to the BS is protected using the primary SA of the SS. Although typically all downstream unicast traffic is protected using the primary SA as well, additionally some selected downstream unicast traffic flows can be protected under static or dynamic SAs. Multicast traffic is protected under static or dynamic SAs (as opposed to a primary SA, which is unique per SS).

13.2.2 PKM

In this section, the PKM protocol mentioned earlier is discussed. An overview of PKM is given in Figure 13.6.

The PKM protocol provides secure distribution of keying material from BS to SS for authorization. The protocol also supports re-authorization and key freshness. As mentioned earlier, PKM is also used to enforce conditional access to network services and thus prevent cloned SSs to access. PKM uses X.509 certificates and two-key triple DES to establish a shared secret—that is, an access key (AK)—between the BS and SS. An AK is used to secure subsequent PKM

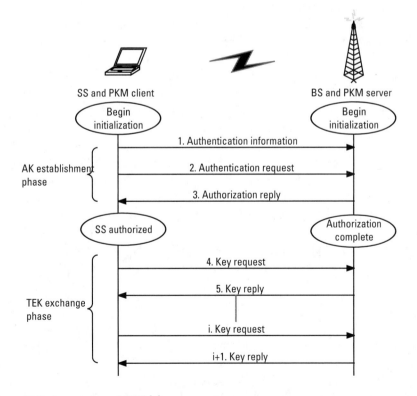

Figure 13.6 An overview of PKM [3].

exchanges of traffic encryption keys (TEKs). The TEK is used to protect data traffic. X.509 digital certificates are used to identify communication parties; the certificate includes information such as device MAC address, public key, serial number, and manufacturer identity. All SSs have factory-installed RSA private/public key pairs or an internal algorithm to generate such pairs, used for PKI. Note that the private key is embedded in the hardware of the SS.

The BS authenticates a SS during the initial authorization exchange. Within the authorization exchange, the SS would then send a copy of this device certificate to the BS. The BS verifies the SS certificate and possibly performs certificate path validation checks. If satisfied, the BS as part of its response to the SS would encrypt the AK assigned to that SS using the public key of the SS. This is the authorization response from the AK; it includes the SAID and the remaining lifetime of the AK.

On reception of the authorization response, the SS proceeds to obtain TEK from the BS. The standard defines a TEK state machine for this purpose. Receiving the authorization response also means setup/finalization of a primary SA and possible static SAs. Having set the SAs, AK, and TEK, the point is to provide key transition using PKM.

A SS should obtain a new AK within the lifetime given; otherwise, it will be considered unauthorized by the BS. A BS has two simultaneously active AKs for a given SS with overlapping lifetime. When the SS sends an authorization request while there is only one AK, the AK transition period is triggered. The new key has a lifetime equal to the remaining lifetime of the old AK plus its own lifetime and has one higher sequence number, modulo-16, than the old AK.

A BS and SS share two active TEKs per SAID. A new TEK is assigned a key sequence number one greater (modulo 4) than that of the older TEK. The BS takes care of initiating the change in TEK. The BS transitions immediately to the new TEK for downlink traffic on the expiration of the old TEK. For the uplink, the BS sends the new TEK, enough in advance, in the key response message, and the transition takes place upon expiration of the old TEK. For encrypting downlink traffic, the BS uses the older TEK; for decrypting uplink traffic, the BS uses the old (if not expired) or new TEK, depending on the information in the header. Thus the BS actually encrypts using a TEK for the second half of its lifetime and will switch to new TEK on receiving a key request message from the SS.

13.3 PKM Security Issues

Several security issues related to PKM are discussed in this section [2, 3]:

- *Key length and incorrect use of cipher mode:* The basic choice of MPDU encryption is DES-CBC with a 56-bit key, while the per-packet IV is computed using an initial IV sent during TEK establishment. This flawed because 56-bit key DES does not provide any meaningful confidentiality protection to MPDUs, and CBC mode requires an unpredictable IV for safe operation. A fixed IV XORed with a sequence number does not meet this requirement.

- *Integrity protection:* The DES-CBC mode for secure encapsulation of MPDUs does not have associated message integrity protection.

- *Mutual authentication:* The PKM protocol provides SS authentication to BS but not vice versa.

- *Key ID fields:* The AK ID is 4 bits in length, and the TEK ID is 2 bits in length. An adversary may replay old messages to trick the SS to encapsulate PKM messages or data MPDUs with old keys. This may allow the adversary to attack the underlying cipher.

- *Replay protection:* The PKM protocol does not protect against replay attacks; neither is there possibility for liveness verification. The lack of replay protection allows an adversary to trick an SS into accepting an old AK as a fresh AK, and this could lead to an attack on the underlying cipher (3DES-ECB) that protects the TEKs.

13.4 PKMv2

IEEE 802.16e focuses on an extended security sublayer with two versions of PKM protocol. PKM version 1 is quite similar to the basic security sublayer, except that it also supports EAP-based authentication. PKM version 2 is meant for mobility and allows preauthentication of mobile SS. The standard also defines a key hierarchy to allow a mobile SS to authenticate itself to the backend AAA server once, irrespective of any number of BSs it may associate with.

13.4.1 Authentication and Access Control

PKM in IEEE 802.16e supports two distinct authentication protocol mechanisms:

- *RSA protocol—PKCS #1 v2.1 with SHA-1(FIPS 186-2):* Support is mandatory in PKMv1 and is optional in PKMv2. In PKMv2, this method supports mutual authentication and authorization.

- *EAP:* This is optional unless specifically required; it is discussed in Chapter 10. This method allows backend authentication.

PKMv2 fixes most, if not all, of the flaws in the original PKM design. In PKMv2, AES-CCM is used as the MPDU encapsulation algorithm. CCM is explained in Chapter 3 and 12.

13.4.1.1 Public Key

PKMv1 public key authorization is explained in Section 13.2.2. The same three messages are used in PKMv2 with modification.

At first, an authorization request message is sent containing a 64-bit MS random number, the MS's X.509 certificate, and a list of cryptographic suites—integrity and encryption algorithms—that the MS supports. The SAID is the SS's primary SAID. This message is not signed by the SS.

The authorization response message from the BS consists of the received 64-bit MS random number, the 64-bit BS random number, and the RSA encrypted 256-bit pre-primary AK (PAK) encrypted with the MS's public key. The message also includes BS certificate; the BS signs the entire message. The MS can verify the signature, and then it verfies the liveness by comparing the received and transmitted MS random numbers. It then extracts the PAK, the associated attributes, and finally the SAIDs. Only the authorized MSS can extract the PAK; therefore, MSS authorization is also verified by this message.

The authorization acknowledgment from the MS consists of the BS random number received in the previous message for liveness proof, as well as the MS MAC address (identity), and includes a cryptographic checksum of the acknowledgment message. The integrity algorithm specified is the one-key CBC MAC (OMAC) algorithm with AES as the base cipher, and the OMAC key is derived from the PAK with 0 as the packet number in the derivation.

13.4.1.2 EAP

Figure 13.7 conceptually depicts end-to-end AAA on 802.16 networks using EAP. The figure borrows terminology from Wi-Fi and is built on the three-party protocol (PKM v2) foundation being defined in 802.16e.

As shown in the figure, over-the-air authentication and encryption (security association) is established using the PKM-EAP protocol. EAP is carried over RADIUS or diameter to the AAA backend. The use of EAP enables support for cryptographically strong key-deriving methods such as EAP-AKA and EAP-MSCHAPv2. Intel also recommends using an end-to-end tunneling protocol such as PEAP or TTLS to afford mutual authentication and 128-bit or better TLS encryption to further enhance end-to-end security (especially in situations where cryptographically weaker EAP methods may be deployed). The AP, APC, or APR serves as the authenticator and hosts a RADIUS or diameter AAA client. All AAA sessions are terminated on an AAA server, which may be in the operator's IP core network or an external IP network in roaming scenarios. The RNSN is merely a conduit for the AAA messages and does not play a significant role in

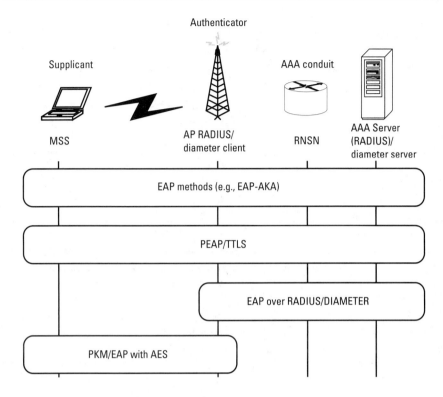

Figure 13.7 E 802.16 security framework [5].

the AAA process. In some instances, the network may employ an AAA aggregator/intermediary but the architecture is not impacted in those cases.

Additionally, the RNSN may host a firewall to filter downstream traffic to a RAN.

13.4.2 Preauthentication

In anticipation of a handover, an MS may seek to use preauthentication to facilitate an accelerated reentry at a particular target BS. Preauthentication results in the establishment of an authorization key (with a unique AK name) in the MS and target BS. The specific mechanism for preauthentication is out of the scope of IEEE 802.16e.

13.4.3 TEK Update

During reauthorization, or when the TEK expires, the SS and the BS do not need to engage in the full RSA or EAP authentication process. Instead, as long as the AK has not expired (and a counter counting the number of three-way exchanges does not reach a configured maximum), the SS and the BS can use the three-way exchange to refresh the TEK. If nonces are not used in the key derivation, note that the KEK and the integrity keys do not change.

References

[1]   *IEEE 802.16-2004 IEEE Standard for Local and Metropolitan Area Networks: Air Interface for Fixed Broadband Wireless Access Systems Part 16,* Piscataway, NJ: IEEE Press, 2004.

[2]   Johnston, D., and J. Walker, "Overview of 802.16 Security," *IEEE Security and Privacy,* May/June 2004, pp. 40–48.

[3]   Hardjono, T., and L. Dondeti, *Secure WLANs and WMANs,* Norwood, MA: Artech House, 2005.

[4]   "Draft IEEE Standard for Local and Metropolitan Area Networks. Air Interface for Fixed and Mobile Broadband Wireless Access Systems: Amendment for Physical and Medium Access Control Layers for Combined Fixed and Mobile Operation in Licensed Bands," IEEE 802.16e Draft 10, August 2005.

[5]   Agis, E., et al., "Global, Interoperable Broadband Wireless: Extending WiMAX Technology to Mobility," *Intel Technology Journal,* Vol. 8, No. 4, August 2004, pp. 173–187.

[6]   Eklund, C., et al., "IEEE Standard 802.16: A Technical Overview of the WirelessMAN™ Air Interface for Broadband Wireless Access," *IEEE Communications Magazine,* June 2002, pp. 98–107.

14

WWAN Security

14.1 GSM Security

GSM is the second generation technology for mobile phone communications. It was accepted as the international standard for digital cellular telephony in the late 1980s, and it started to be deployed in the 1990s. GSM replaced first generation cellular phone systems, which were analog systems that could support only a limited number of users. Two of the major systems that were in existence were the advanced mobile phone system (AMPS), the standard chosen in the United States, and total access communications system (TACS), mainly deployed in Europe. The downfall of first generation systems was the need for greater capacity as well as a technology that could support international communications. Static and cross-channel interference are major annoyances with analog phones while nonexistent with digital. Last but not least, security and privacy can be easily implemented on digital networks through encryption methods.

Today GSM is one of the most widely deployed digital cellular telephone systems in the world. Competing technologies are the United States–developed CDMA and time division multiple access (TDMA). Although these technologies are intrinsically incompatible, many phones today support multiple technologies, and mobile telephone companies have made agreements to allow users to call and be reached independently of the service offered in their coverage area.

When GSM was conceived and standardized by the European Telecommunications Standards Institute (ETSI), two security services were targeted: authentication and encryption. Since the goal for authentication was to allow the telephone company to identify the user for billing purposes, only one-way authentication was requested, whereas trust in the network was considered implicit. Encryption was designed to protect the air link between the mobile user and the telephone operator ground antenna, while communication confidentiality within the operator's network was not taken into account.

GSM security relies on symmetric key cryptography, a secret key Ki is shared between the user and the network operator. This key is the secret used to perform the authentication protocol and to calculate session encryption keys. If the value of this key is revealed, an attacker may impersonate a victim as well as eavesdrop on his conversations; for this reason, the standardization committee decided to store this key in a smart card, which is a tamper-resistant device. Smart cards used in GSM are called SIM cards. Besides storing the secret key Ki, the SIM card also provides a protected environment within which sensitive cryptographic operations are performed. Every SIM card is personalized (i.e., it contains the unique user identification IMSI code and secret key Ki). A SIM card can be moved from one handset to another without the user having to change his subscription contract or his telephone number.

Another service offered by GSM is user anonymity for privacy protection. An IMSI is a nonconfidential value linked to a particular user. IMSI knowledge would allow identifying the user's physical location worldwide, while use of a temporary identity, known as temporary mobile subscriber identity (TMSI), can provide anonymity. The TMSI is frequently updated (every time the user moves to a new location area or after a certain time period) to avoid linking user information with TMSI. There are situations where IMSI use is mandatory (e.g., on the first use of the mobile after purchase, on the first use of the mobile under the coverage of another operator, or whenever the provided TMSI cannot allow establishment of user identity).

ETSI defined three algorithms to achieve authentication, encryption, and encryption key derivation from Ki. Input and output lengths as well as key lengths were specified. Algorithm choice for authentication was left open, whereas encryption algorithms must be standard to allow for roaming between operators.

Reference algorithms were designed by ETSI and kept secret from the public for operator use only; they are discussed in Chapter 3. A3 allows calculating the response $SRES$ to a challenge $RAND$ sent by the network operator to the user. A8 uses the same $RAND$ as input to calculate the encryption key Kc. A5 is the voice encryption algorithm. Algorithms A3 and A8 are often combined in a single algorithm referred to as A3/A8, the use of which is to calculate the challenge response $SRES$ and encryption key Kc given the secret key Ki and the random input $RAND$. Besides reference A3/A8 algorithms, operators have the option to implement proprietary algorithms, or published algorithms that fit the requested characteristics. The mobile phone and the visited network must support the same A5 algorithm to allow encrypted voice communication.

GSM security is based on triplets, including

- The challenge *RAND;*
- The challenge response *SRES;*
- The voice encryption key *Kc.*

The algorithms to calculate *SRES* and *Kc* from *RAND* and the secret key *Ki* are described in the following sections.

14.1.1 User Authentication

User authentication is achieved by performing a challenge response between the network and the user. (See Figure 14.1.) To be more specific, user authentication occurs in the SIM card whereas network authentication occurs in the authentication center (AuC) or the home local register (HLR). If the mobile user is in a visited local register (VLR) coverage area (i.e., an area under the coverage of an operator that has roaming agreements with the operator the user subscribed to), the AuC or HLR transfer the authentication results, success or failure, to the VLR.

Network authentication is not required in GSM, as the cost to build a rogue network station was considered sufficiently prohibitive to put off potential attackers.

The algorithm A3 is implemented in the SIM card and the AuC or HLR. To authenticate, the user receives a 128 bit random challenge *RAND* from the network. Using the 128-bit secret key *Ki* algorithm, A3 computes a 32-bit challenge response signed response (SRES) and transmits it to the network for verification.

The authentication procedure is outlined in the following steps:

1. Authentication is initiated by the user whenever he wants to make a call from his mobile (MS) or go on standby to receive calls. The user transmits his identity through TMSI and the authentication request.

2. The network establishes the identity of the SIM through the 5-digit TMSI. If the TMSI is recognized, the VLR sends a request for authentication to the HLR; if not, it will request the user's IMSI.

3. The HLR generates a 128-bit random *RAND* challenge. Using the user K_i and *RAND*, it applies A3 and calculates the expected $SRES_{HLR}$. *RAND* and $SRES_{HLR}$ are both sent to the VLR.

4. The VLR sends *RAND* to the SIM.

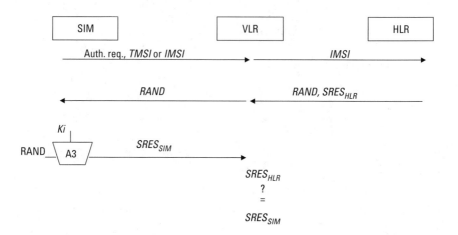

Figure 14.1 GSM authentication.

5. The SIM calculates the $SRES_{SIM}$, using its secret key K_i and the challenge $RAND$. $SRES_{SIM}$ is sent to the VLR for verification.

6. If $SRES_{HLR} = SRES_{SIM}$, then the SIM is authenticated and allowed access to the network. If $SRES_{HLR} \neq SRES_{SIM}$, then an authentication rejected signal is sent to the SIM and access to the network is denied.

14.1.2 Voice Encryption

The frequencies over which voice is transmitted are public, so voice encryption is necessary to avoid interception of the signal over the air. Once the signal reaches the operator's BS, it will be transmitted to the receiver over a wired or wireless mean. In either case, ETSI didn't define any protection: voice transmission in clear over a wired means is publicly accepted, as this is what happens for fixed base telephone conversations, and voice transmission in clear over wireless portions of the network is supposedly not at risk, as it is assumed that the attacker is not aware of the communication path within the operator's network. Voice will only be encrypted from the BS to the receiver if the receiver is herself a mobile user. In the latter case, it should be noted that a different encryption key will be used between the caller and his base station and between the receiver and her base station.

Voice encryption is not mandatory; the choice whether or not to accept an unprotected communication is up to the network. A session encryption key must be computed before a secure communication can take place. The encryption key Kc will change after each user authentication, since the same $RAND$ value is used for encryption key derivation.

The algorithm A8 is implemented in the SIM card and the AuC or HLR. To generate the 64-bit encryption key K_c, the SIM uses the 128-bit random challenge *RAND* from the network and the secret key K_i. K_c is transmitted to the MS for voice encryption.

The voice encryption algorithm implemented in a MS is A5. It's a stream cipher that takes K_c as input and produces a key stream as output. Ciphertext is obtained by XORing the plaintext and the key stream.

Multiple versions of the A5 algorithm have been defined; the network and MS must support at least one common version to communicate securely. The most widely used A5 algorithm today is A5/3 (A5 version 3); it is based on Kasumi and described in Chapter 3.

When a MS wishes to establish a connection with the network, it indicates which version of the A5 algorithm it supports. If the MS and the network have no versions of the A5 algorithm in common, the network decides whether to accept an unciphered connection or to release the connection. If the MS and the network have at least one version of the A5 algorithm in common, then the network selects the one of its choice.

The voice encryption procedure is outlined in the following steps:

1. The SIM card applies algorithm A8 to the 128-bit input *RAND* using key *Ki* to calculate the encryption key *Kc*.

2. The SIM card transfers the encryption key *Kc* to the MS.

3. When the MS wants to establish a connection, it informs the network of the A5 algorithms it supports.

4. If the MS and the network have at least one version of the A5 algorithm in common, then the network selects the one of its choice.

5. If the MS and the network have no A5 algorithms in common, the network accepts an unciphered connection or releases the connection.

14.1.3 Other Security Features

SIM authentication and voice encryption are considered GSM main security features, but a number of minor protections are also available. We have already mentioned anonymity, which allows the concealment of the SIM's permanent identity IMSI, linked to a particular user's identity. The same TMSI should not be used for a long time period to avoid user traceability.

IMEI aims to reduce mobile phone theft. The network can request the IMEI of the mobile station it's communicating with. If the value provided corresponds to the IMEI of a stolen phone, the network may interrupt the communication. Unfortunately, no security feature protects IMEI integrity, so the barring of stolen phones depends on the terminal providing the genuine IMEI to the network.

User-to-SIM authentication may be requested before a user is allowed to employ SIM services. This proof, whose goal is to limit the use of stolen SIM cards, is generally accomplished by PIN verification. PINs are generally 4–16 digits long, but users can disable this feature.

Also, to limit the use of a stolen mobile platform with a different SIM card, mobile phone owners can pair their device with their SIM card. This feature, known as SIM lock, allows a SIM card and a mobile platform to share a secret. The SIM will be denied access to the terminal unless it can prove knowledge of the secret.

14.1.4 Security Limitations and Attacks on GSM

A number of security limitations have been reproached to GSM. The most obvious is its lack of support for mutual authentication, which enables an adversary to set up a false BS and communicate with any user, since only user authentication is requested and since the network can opt for nonencrypted communications.

As we've mentioned in previous sections, data within the network is not protected. This concerns voice, which is transmitted in clear, as well as signaling information, including cipher keys and authentication tokens. Any adversary that can access an operator's network from the inside will be able to impersonate victims or network elements as well as eavesdrop on communications.

Yet another GSM design limitation is its lack of integrity protection. This is not a major issue on voice communications, where throughput is more important than error detection or protection and where personal voice characteristics allow us to recognize who we are speaking to. Lack of integrity becomes an issue in GPRS, a technology based on GSM, where data transmission is supported.

On top of design limitations, many GSM algorithms have been broken over the years. First, the variable and key lengths are too short to be considered secure, given the increase in computational power since the definition of GSM. The secret key K_i is 128 bits long, which is still acceptable today in symmetric key cryptography, but SRES is only 32 bits long, giving a 2^{16} chance of collision using the birthday paradox, and the encryption key K_c is only 64 bits long. Even worse, in early versions of A5 algorithm, only 54 of the available 64 bits were used for encryption.

Particular implementations of the A3/A8 authentication and cipher key–generation algorithms, as well as of the A5 algorithm have been breached. The first reference version of the A3/A8 algorithm designed for mobile telecommunication operators, called COMP128 v1, was broken in the early 1990s. COMP128 v1 was kept secret and was not publicly revised. Once it leaked, it was attacked and broken. The first attacks by Berkeley students in 1992 showed that by analyzing COMP128 v1 output on chosen *RAND* values, the secret key K_i could be retrieved. Once K_i is known, an attacker can clone a SIM card and impersonate a user or make calls at the expense of the victim whose SIM card was cloned. Later COMP128 v1 was also broken by side channel attacks based on power consumption [1].

Early versions of the A5 voice encryption algorithm have also been reverse engineered and broken. The A5 version 1 algorithm was broken in 1994 by an attack that allows finding the voice encryption key by eavesdropping on a two-minute conversation [2]. In 1999, an attack on the weaker A5 version 2 was announced [3]. Another attack on A5 version 2 based on a ciphertext_only analysis of encrypted off-the-air traffic was published in [4].

Elad Barkhan, Eli Biham, and Nathan Keller [5] have shown a ciphertext-only attack against A5/2 that requires only a few dozen milliseconds of encrypted off-the-air traffic. They also extended their attack against A5/1 and A5/3 on mobile phones that support A5/2 by retrieving the key first used in an A5/2 algorithm and then switching to another A5 version.

GSM networks lack the flexibility to quickly upgrade once security breaches are identified. In Chapter 3, we describe the encryption algorithm A5/3 and the authentication and key generation algorithm MILENAGE, but these have not been widely adopted in GSM.

14.2 3GPP Security

The 3GPP Agreement was signed in 1998 to complete a set of globally applicable technical specifications for a 3G mobile system based on the evolved GSM core networks and the radio access technologies based on UMTS terrestrial radio Access. A separate standardization body, 3GPP2, is developing another third generation mobile cellular system based on CDMA2000 and an evolution of the North American standard ANSI-41.

3GPP security specifications describe both access security and network security. Access security is improved by adding services not provided by GSM and correcting GSM vulnerabilities by employing different algorithms. Network security is an entirely new feature compared to GSM.

3GPP provides over-the-air mutual authentication between the user universal subscriber identity module (USIM) and the network, encryption, and integrity of user and signaling data.

Since 1998, the 3GPP technology has been evolving, and multiple releases of the specification have been published. The first release of 3GPP specifications, release 99 [6], was essentially a consolidation of the underlying GSM specifications and the development of the new UMTS Terrestrial Radio Access Network (UTRAN). Innovative services defined include multimedia messaging service to send text, audio, images and video clips, location services to send user's emergency and commercial data according to their location, mobile station execution environment to allow a mobile station to negotiate its execution environment, and access to the Internet or an ISP. In Release 4 [7], major security enhancements concern the definition of encryption algorithms based on Kasumi and the establishment of mobile application part (MAP) application layer security.

The main improvement in release 5 [8] is the ability to support IP-based communication between network elements. Confidentiality, integrity, authentication, and antireplay protection are obtained thanks to IPSec. Release 6 is now finalized.

14.2.1 3GPP Authentication and Key Agreement

3GPP provides mutual authentication and key agreement between the user USIM and the network through the AKA protocol. (See Figure 14.2.) AKA is a secret key algorithm; a secret key K must be shared between the USIM and the HLR. It is the HLR that generates authentication values and transfers them to the VLR of the network under which coverage the user is located the moment authentication is performed.

Authentication is requested by the USIM. Once the HLR has transferred authentication values to the VLR, exchanges occur between the USIM and the VLR. Authentication values consist of a quintet (in analogy to GSM triplets) including:

- The challenge *RAND;*

- The challenge response *XRES;*

- The cipher key *CK;*

- The integrity key *IK;*

- The authentication token *AUTN.*

The AKA procedure is outlined in the following steps:

1. Authentication is initiated by the user whenever he wants to make a call from his mobile station or go on standby to receive calls. The user transmits his identity through TMSI and the authentication request.
2. The network establishes the identity of the USIM through the 5-digit TMSI. If the TMSI is recognized, the VLR sends a request for authentication to the HLR; if not, it will request the user's IMSI.

3. The HLR generates the AKA quintet $Q = (RAND, XRES, CK, IK, AUTN)$ and sends it to the VLR.

4. The VLR sends *RAND* and *AUTN* to the USIM.

5. The USIM verifies if *AUTN* is acceptable, where *AUTN* is the network authentication token. If *AUTN* is valid, the USIM calculates the expected response $XSRES_{USIM}$ using its secret key K and the challenge *RAND*. $XSRES_{USIM}$ is sent to the VLR for verification. The USIM also calculates the encryption key CK and the integrity key IK.

6. If $XRES_{HLR} = XSRES_{USIM}$, then the USIM is authenticated and allowed access to the network. If $XRES_{HLR} \neq XSRES_{USIM}$, then an authentication rejected signal is sent to the USIM and access to the network is denied.

The authentication algorithm was not standardized by the 3GPP organization because the architecture demands that every operator manages her users' authentication and sends the authentication quintet Q to the VLR. Nevertheless, the reference algorithm MILENAGE, described in Chapter 3, was designed and is used by most operators.

AKA was designed in such a way as to facilitate roaming and handover between 3GPP and GSM networks because it is expected that for a long transition period, both networks will coexist. To ease roaming between 3GPP and 3GPP2 networks, 3GPP2 has decided to adopt AKA as its authentication and key agreement scheme as well.

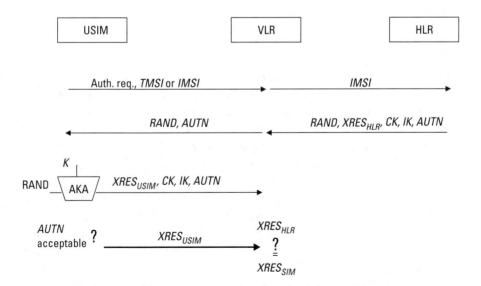

Figure 14.2 3GPP authentication.

14.2.2 3GPP Encryption and Integrity Functions

For compatibility between operators for roaming purposes, encryption and integrity algorithms have been fully standardized [9]; these algorithms are described in Chapter 3.

User and signaling data confidentiality is suggested but not mandatory. At present, there is only one standard encryption algorithm that can be used, the f8 function based on KASUMI block cipher. The $f8$ function calculates a keystream based on the 128-bit key *CK,* a time-dependent variable *COUNT,* a bearer identity *BEARER,* the transmission *DIRECTION,* and the required key stream *LENGTH.* The ciphertext is obtained by XORing the plaintext and the keystream.

Signaling data integrity is mandatory to prevent a number of attacks, including replay attacks, man-in-the-middle attacks, and security downgrading when multiple security algorithms are supported. Data integrity allows detection of data modifications as well as identification of the message sender. At present there is only one standard integrity algorithm that can be used, the $f9$ function based on KASUMI block cipher. The $f9$ function calculates a MAC based on the 128-bit key *IK,* a sequence number *COUNT,* a random value *FRESH* that guarantees the communication freshness, the transmission *DIRECTION,* and the input *MESSAGE.* A 32-bit MAC is appended to all communications over the radio link.

User data integrity protection is not specified, but the use of sequence number in message transmission allows detection of whether user messages are deleted or inserted.

14.2.3 3GPP Network Security

Signaling system number 7 (SS7), defined by the International Telecommunications Union (ITU) for the PSTN in the 1980s, was the first de facto standard to be used for communication within and between operators' networks. No standard security means are defined in SS7, since wired telephone operators rely on the private nature of the network to infer that attacks will be limited. To adapt SS7 to wireless communications, the MAP protocol was developed and included in release 99. Wireless access to operators' networks nevertheless implies new breaches for attackers, so 3GPP developed security mechanisms specific to MAP (MAPsec) in release 4. MAPsec has been improved throughout 3GPP releases.

Unfortunately, MAPsec provides some degree of protection only on the mobile part of the signaling protocol, not on the entire SS7 protocol. Instead of defining a security protocol for SS7, more and more operators are now switching to IP and IPsec for security. Moreover, MAP can run on top of IP, leaving the choice between IPsec and MAPsec for security.

14.2.3.1 MAPsec

MAPsec is an application layer security protocol, fully useful if applied by all interconnected operators.

Before protection can be applied, SA must be established between the respective MAP network elements. SAs define, among other things, which keys, algorithms, and protection profiles to use to protect MAP signaling. Network operators negotiate among each other and distribute to all network elements the necessary MAPsec-SAs to use between networks.

Each SA contains the sending and receiving public land mobile network (PLMN) identifier, a SPI to identify the SA, an integrity and encryption key and the respective algorithms to use, a protection profile identifier (to identify the security features provided), and an expiration date for the SA.

An interdomain SA and key management agreement should

- Define how to carry out the initial exchange of MAPsec SAs;
- Define how to renew the MAPsec SAs;
- Define how to withdraw MAPsec SAs;
- Decide if fallback to unprotected mode is to be allowed;
- Decide on key lengths, algorithms, protection profiles, SA expiration times, and so forth.

The security services provided by MAPsec are

- Data integrity;
- Data origin authentication;
- Antireplay protection;
- Confidentiality (optional).

MAPsec provides three different protection modes:

- *Protection mode 0:* No protection;
- *Protection mode 1:* integrity, authenticity;
- *Protection mode 2:* confidentiality, integrity, and authenticity.

MAP messages protected by means of MAPsec consist of a security header and the protected payload. In all three protection modes, the security header is transmitted in cleartext. The protected payload format is described in [10].

At present, the only mandatory algorithms standardized in MAPsec are AES in counter mode with 128-bit key length for encryption and AES in a CBC MAC mode with a 128-bit key for integrity.

14.2.3.2 IPsec

The security protocols to be used at the network layer to protect IP signaling traffic are the IETF-defined IPsec protocols, a description of which is provided in Chapter 8. In [11], 3GPP defined a minimum set of features required for interworking purposes. IPsec is restricted to ESP and tunnel mode only. Also, key management and distribution between security gateways, defined next, is handled by the protocol IKE. Within their own network, operators are free to use any IPsec feature, including the ones not incorporated in [11].

3GPP defined security domain for network protection (i.e., a network in which the same level of security and usage of security services is provided). Typically a network operated by a single operator will constitute one security domain. Security gateways are entities on the borders of the IP security domains and will be used for securing native IP-based protocols. All IP traffic shall pass through a SEG before entering or leaving the security domain.

The security services provided by IPsec are

- Data integrity;
- Data origin authentication;
- Antireplay protection;
- Confidentiality (optional);
- Limited protection against traffic flow analysis when confidentiality is applied.

At present, the only IPsec algorithms mandatory in [11] are 3DES and AES-CBC with 128-bit keys for encryption, and HMAC_MD5 and HMAC_SHA1 for integrity and data origin authentication.

References

[1]   Benoit, O., et al., "Mobile Terminal Security," Cryptology e-Print report, 2004/158.

[2]   Biryukov, A., A. Shamir, and D. Wagner, "A Real Time Cryptanalysis of A5/1 on a PC," *Fast Software Encryption Workshop 2000,* 2000, New York, April 10–12, 2000.

[3]   Goldberg, I., and D. Wagner, "Rump session of Crypto 99," Santa Barbara, CA, August 15–19, 1999.

[4]   Boman, K., et al., "UMTS Security," *Electronics and Communication Engineering Journal,* October 2002, Vol 14, Iss. 5, pp. 191–204.

[5] Barkan, E., E. Biham, and N. Keller, "Instant Ciphertext Only Cryptanalysis of GSM Encrypted Communication," *Crypt 2003,* Santa Barbara, CA, August 17–21, 2003.

[6] Overview of 3GPP release 99, Summary of All Release 99 Features, ETSI Mobile Competence Centre, July 2004, http://www.3gpp.org/ftp/Information/WORK_PLAN/Description_Releases/Rel99_features_v2004_07_20.zip.

[7] Overview of 3GPP Release 4, Summary of All Release 4 Features, ETSI Mobile Competence Centre, v 1.1.0 (draft), http://www.3gpp.org/ftp/Information/WORK_PLAN/Description_Releases/Rel4%20features_v_2004_07_16.zip.

[8] Overview of 3GPP Release 5, Summary of All Release 5 Features, ETSI Mobile Competence Centre, September 9, 2003, http://www.3gpp.org/ftp/Information/WORK_PLAN/Description_Releases/Rel5_features_v_2003_09_09.zip.

[9] 3GPP TS 35.201, 3rd Generation Partnership Project; Technical Specification Group Services and System Aspects; 3G Security; Specification of the 3GPP Confidentiality and Integrity Algorithms; Document 1: f8 and f9 Specification (release 6).

[10] 3GPP TS 33.200, 3rd Generation Partnership Project; Technical Specification Group Services and System Aspects; 3G Security; Network Domain Security; MAP application layer security (release 6) V6.1.0 (2005-03).

[11] 3GPP TS 33.210, 3rd Generation Partnership Project; Technical Specification Group Services and System Aspects; 3G Security; Network Domain Security; IP network layer security (release 6) V6.5.0 (2004-06).

15

Future Security Challenges*

The future of telecommunications is to reach mass population in all regions of the world. Telecommunications and its services will become part of life, as is breathing to mankind. Fortunately, for whatever we develop, there is always a next step to it. After all, that keeps our world going. Currently we are again at that stage of work on future generation communications where these words have not yet achieved a consensus. In this paper let us look at the crystal ball and try to materialize at least in words what we see in it [1–21].

In order to discuss future security challenges, we need to understand what the future is so we first define fourth generation (4G) communication system and beyond 3G (B3G). Then we look at the requirements for future generation communications from the perspective of the users, the operators, and the service providers. Next, the technologies that should and are being developed to materialize the future generation are discussed. A dip is also taken into the ongoing standardization or prestandardization efforts. At first, the introduction section tries to raise questions on the "future," and the chapter ends with security issues that need to be tackled.

15.1 Introduction

Telecommunications is in its infancy; we have said it [1] and heard it, but what does it entail?

---

* Parts of this chapter are from Chapter 8, "Future Generation Communications," in *802.11 WLANs and IP Networking: Security, QoS, and Mobility* by Anand R. Prasad and Neeli R. Prasad, Artech House, 2005.

Let us step back and look at what telecommunications provides. At first with wired telephones, people could communicate without letters or visits to each other. This was the start of voice communications; next came analog mobile phones, which made this voice mobile, and then the digital era increased the capacity. 3G communications came next and stumbled before even taking the first step. The reason for this stumble was that 3G was aiming at services other than voice communications (this being said with eyes closed to trouble it caused for the telecom industry, the spectrum auction in Europe, and financial mishaps). What does it mean? The infant is growing? What shall we do? Which path should we take?

The other side was the boom of the Internet, which opened a new world and new ways of communications (e.g., e-mailing, peer to peer) and made a plethora of data available to us. WLANs came in and made the Internet wireless and mobile. WLANs grew at an amazing pace with a short product lifecycle and required continued standardization effort toward improving the capabilities and functionalities of the technology. WLANs have been successful exactly where mobile communications, to a reasonable extent, failed. Now voice over WLAN is the vision, combined with mobility; in this arena mobile communications has already set a high standard. VoIP, although gaining momentum now, has been stumbling for a long time, and as yet it has no mobility. WLAN deployments have a small footprint, and mobility is still being worked on. Seamless mobility for WLAN will be ever complex. These points from mobile systems and the WLAN side are represented in Figure 15.1. So what to do? How to proceed? Not only technologywise but market acceptancewise.

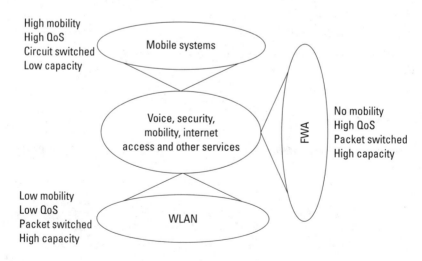

Figure 15.1 Technologies moving toward a common goal.

Other standardization work is going on. 3G standards are giving solutions for interworking with WLAN and other technologies. IEEE 802.16 and IEEE 802.20 are looking at mobility with high bandwidth, and IEEE 802.21, like 3G, is also looking at interworking and thus handover between various technologies. There is another word looming in the air, and that is known as 4G; with it is an aura, an aura that creates confusion, the aura is also known as the B3G. Now what will happen in the future? What are these different Gs and interworking? How do we step forward? How should we approach the market? What technologies should we develop? How can telecommunications become such an integrated part of human life that breathing will become its synonym?

In this paper we try to tackle most of these questions. We look into what 4G, B3G, and—to add to the confusion—next or future generation communications will be.

15.2 The Need for Future Generation Communications

Let us first have a look at the need for future generation communications [2]. Note that we have evaluated none of the information given in this section; text represents the opinion of the author or information taken from the references.

15.2.1 What Will Sell?

One can never say what the user will want in the future. Whatever you say will be way off the reality. So what can one do? Learn from past. This is the logical tactic mostly used, but we can go way off the mark using this tactic when it comes to such futuristic predictions [3]; 3G proves this logic. Nevertheless, voice is the best bet for a future. What else? Well, person-to-person communication in any possible way is the service that really sells, seamless service provisioning being a part of it. Next is machine-to-machine communications; this will pick up strongly in the future. In general, a product is bought (or sold) if any of the following is fulfilled:

- The product solves a problem being faced by the customer (e.g., the Intuit software by Scott Cook that provided an easy way to fill out tax forms [4]).

- There is a hole in the market for a given product (e.g., the WCDMA test equipment developed by Quintillion Technologies from Japan) [5].

- The product is cheaper than others in the market and is within a quality limit (e.g., fixed and mobile operator's voice service cost in the United States).

Besides these, there are several reasons a product sells, such as brand name. Here, however, we will try to find the need of the user based on these three principles.

15.2.2 Is It Common Sense?

Let us look at two examples that defy the conventional wisdom of what will or should sell.

The first example is Telenor's electronic shepherd project. The shepherds with sheep grazing in forests needed some way of locating lost sheep or at least determining whether the sheep were alive and so they could find them. In response to this request, Telenor developed a solution where each sheep has a radio connected to it; these sheep create an ad hoc network with the mother sheep as the central controller. The information about each sheep can thus be accessed by the shepherd. Thus Telenor has expanded its market out of the already saturated market of mobile services in Norway.

Another example is i-mode. It is said very often that i-mode or other mobile Internet services in Japan made it because of the lack of space in Japanese houses for computers. The fact is that Japan is good at miniaturizing electronic goods, and Japanese people like new electronic goods, including computers, so the lack of space is simply a nonissue. Let us look at it differently. Until almost the end of the 1990s Japan was lagging behind in Internet connectivity, so the Japanese government started promoting broadband access in the country; a major share of work was Japan gigabit network. Now i-mode had come before that and naturally captured the market. Today in Japan one can get xDSL as high as 48 Mbps for barely $10; this might be the cheapest in the world. Anyway, getting back to the olden days, another reason for i-mode's success was the improved voice quality; i-mode terminals use full-rate instead of half-rate speech codec as in standard personal digital cellular terminals. Higher speech quality leads to longer conversations, and thus the beginning of i-mode showed growing revenue for voice. Other i-mode services, such as searches for restaurants, also led to extra voice revenues (i.e., one finds a good restaurant through i-mode and calls 10 friends). Only later did a new service provision increase the data revenue, with services such as gaming, photo or short video clips, messaging, location services, and ring tone download. Japan had found a new source of revenue, and thus the price war over voice services that occurred in the United States did not happen; operators were competing on service types. The increased usage of networks led to the lack of spectrum, and thus the need for 3G systems came to Japan (even 3G spectrum might not be sufficient). So one would say the operators had achieved their goal: maximize the use of networks or increase the traffic on their network. Now let us look at the other side; in 3G networks, flat-rate services have started in Japan. This now changes the whole concept; now the operators should want users to decrease their access to the network. The goal of the operator as always will be to keep as many users as possible, create and improve customer loyalty, attract new customers, and decrease the churn rate. However the operator would prefer to decrease the traffic on the network simply because an increase in traffic now does not increase revenue because of flat rates. At the same time, operators will have to keep their services attractive for the users. So now where should the operators go? They have to find a way to keep the customers and increase revenue with attractive services not requiring as much use of the network. In near future, the Japanese market will see several changes due to number portability, mobile WiMax introduction with new licenses, new mobile operators, and IP multimedia subsystem (IMS) introduction.

One could think that current Japanese situation will mean that solutions should be developed to hand over users to cheaper networks while keeping the expensive network for premium users or services. Further, this implies the availability of services close to the users—thus, the necessity of mesh or ad hoc networks. Newer services should be developed for which users will pay additional charges on top of the flat rate for access.

This section has given views from two totally different markets for mobile communications. It should be noted that the business models in Europe and Japan are different. In Europe, the customers usually stick to the brand name of the vendor, as the standard defines all the details. On the other hand, in Japan, the customers stick to the operator, as the standard gives space for operators to develop new services or solutions. So we see the operator being dependent on the vendor in Europe and other way around in Japan. Although recent trends in no-brand terminal developers (or original design manufacturers, e.g., Ben Q.S. Arima) with OSs like Microsoft is slowly changing the ballgame in Europe, too; terminals are available in Europe with operator brand.

15.2.3 How to Know What Will Sell

There is a Hindi saying, "Koop Mandook," meaning a frog from a well. Similarly mobile communications was in a well, the well of voice service. Mobile communications came out of the well in the form of 3G, and WLAN is moving toward QoS-based services. 3G stumbled; WLAN—not as visibly due to lower costs—is also stumbling, but stumbling is good for the lessons it gives. How does one learn from the lessons? The industry has learned to become *user centric*; that is good but its meaning is not understood. The way out is to ask the user, but even what should be asked is unclear. Answering questions about what service they'll need and which products should appear in 10 years is surely difficult for users. The trick is to see how the user uses a product and maybe even lives her everyday life; this will give us a better idea about the possible needs of the user and thus the product. The amazing fact is that this line of thought brings us to a junction where art and technology seem to meet. E. H. Gombrich in his famous book [6] discusses how people see things and perceive them. He says that people should learn how to see, that artists see more, and we appreciate a piece of art if it is close to the nature that is visible to us and we can thus associate with it. This way of seeing the customer is something the mobile business should pay attention to. Another side is the work by C. M. Christensen [3], which says one should try out new products and then learn from the sales and marketing experience. Lessons learned should tell the team what to do to make the product successful, keeping in mind that the initial targeted market might not ultimately use it. This method of Christensen provides a way for the industry to see, as Gombrich said, the mobile market.

The discussion in the previous paragraph also brings us to the junction where we have to say that focusing on one thing can lead to single-mindedness. The point is also to have a broad vision and simple thought. So what does this mean? Let us have a sneak preview. In most developing countries, the society is getting old and thus services like telemedicine and, due to lack of teachers, teleeducation would make sense. On the other hand, for fast-developing countries like India, where more than 50 percent of the population is under 20 years of age, a different approach will be required. Of course here we have not talked about culture and the effect of globalization at all.

Another line of thought is to learn how children use things and behave. This has been proposed by several people. After all, something that will come in the future will eventually be used by the children.

Let us see a few examples based on observations: Barely at the age of one, my daughter used to pick up small toys that would fit in her hand and try to mimic a phone conversation. This shows the need of connectivity, and the solution is ubiquitous communications with context awareness. Another example happened just a day or two back during a German lesson, when I found out that people hardly remembered what a turntable is; today, kids only know mp3 players. So it is also the technology the future generation is interfacing with. Objects such as mobile phones, mp3 players and DVDs (already DVDs are being replaced by hard disk drives) are normal for them, at least in the developed world.

We should also observe things in our own daily life; for example, I do not like the phone ringing during dinner. It should be possible for the phone to know that I am having dinner and allow only emergency calls or calls from people of high priority in my priority list to come in.

15.2.4 Different Perspectives

Let us now look at future generation needs from the point of view of the user, the operator, and the vendor. Many of these needs have appeared in an earlier text in one form or another. Although the term *user* is used in this text, it should be noted that the subscriber or someone else could be the user. The point is that whoever the subscriber is, the user is the one using the service at the end and thus holds the right to decide the fate.

Looking from the user perspective, it is natural to have a reasonably cheap and easy to use solution. The solution from the user perspective should allow access to services at all times.

The vendor, on the other hand, will look for solutions that will be easy to implement and such that one design is reusable for different products. Similar to the user, the vendor will want the solution to be cheap to maximize profit. The solution should be such that it minimizes implementation errors. The vendor will have to implement the mobile platform such that it is easy to use.

From the operator point of view, once again the main thing that comes out is the reduction of the cost: the cost of network elements, the cost of deployment, and finally the cost of operations and management of the network. These are just a few standard points. The operator will provide the services that the user will need and will be easy to use. In future generations, the operator will have to communicate with items and businesses in daily usage by users (microwave ovens, doorbells, the home entertainment companies, consumer electronics, the auto industry, and so forth; see Figure 15.2). Note that we do not say DVD player, VCR, or a particular music system because they will not survive the coming 10 years.

15.3 Defining the Future

Each wireless technology is moving toward future standardization. The standardization work is mainly focusing on wireless IP-based QoS provision for any type of data, where data is everything, be it voice, video, or Internet access. All these standards are, mostly and sensibly for the first time, looking at security from the beginning instead of filling the holes as it develops. In this section, the terms B3G and 4G are defined.

Since 3G did not launch as it was envisaged, using the term 4G had become a taboo. It seems this taboo gave birth to the term B3G, which should have been an underground synonym for 4G but now has developed its own meaning. There are different views on the definition of 4G; some say that any technology that provides data rates above 100 Mbps is 4G. In the following, our definitions, which are generally accepted in Europe, of B3G and 4G are given.

The International Telecommunications Union—Radio Communications' (ITU-R's) vision calls for interworking/integration of available technologies and development of a new air interface that would work at 100 Mbps. The new air interface is defined as 4G and the interworking/integration of technologies is

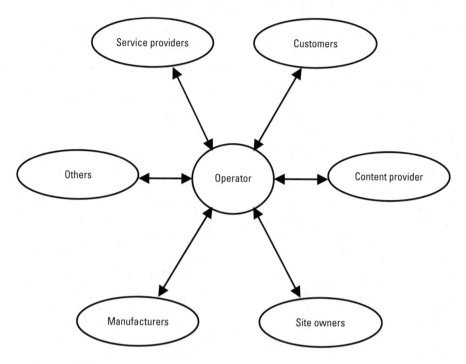

Figure 15.2 Operator's role in the future.

defined as B3G, which also includes 4G air interface (see Figure 15.3). One can see that standardization work of each technology is mainly focusing on wireless IP-based QoS provision for any type of data. Here, data is everything, be it audio, video, games, or any other application. Basically this means an integration of services. All these technologies overlap each other's areas of services to some extent. This is illustrated in Figure 15.4. (WMAN is not shown.)

Thus a move toward integration of technology is a logical next step to provide service continuity and higher user experience (quality of experience). (See Figure 15.5.)

Today there are several operators or stakeholders for different types of technologies and networks; in B3G era there will be a need for handover between them. A common layer for all these technologies and stakeholders will be IP but IPv4 as we know is fragmented due to the lack of addresses. Even though IPv6 is available, v4 is here to stay. All this together forms B3G and with the variety come several technical issues.

As any new system takes about 10 years to develop and deploy (see Figure 15.6), work on B3G and 4G has already started. 3G has gone through trouble, although recent news show that 3G services are expanding; lessons should be learned from it while developing 4G.

To be precise, 4G basically gives a higher data rate with mobility, while B3G gives continuous connectivity. The generation beyond that will give the user real ubiquity. The future generation will provide ubiquitous

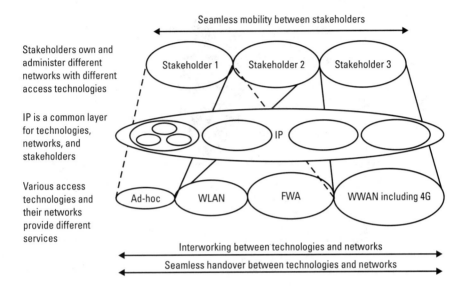

Figure 15.3 Defining B3G.

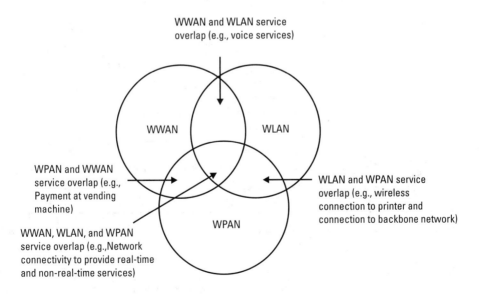

Figure 15.4 WWAN, WPAN, and WLAN overlap.

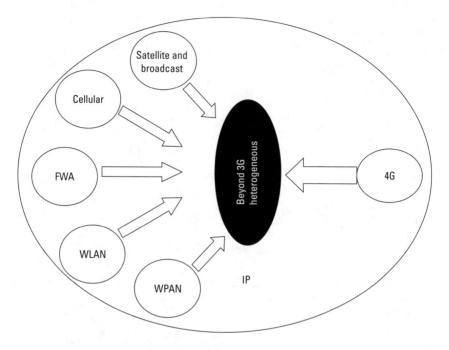

Figure 15.5 Future of telecommunications.

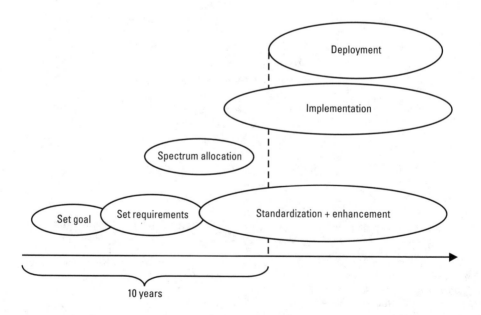

Figure 15.6 Time required for new technology development and deployment.

communication through ubiquitous networking and thus bring ubiquitous services to the user.

The development toward B3G will thus mean that the backbone network used today by the mobile operators will be completely changed and replaced by IP. There will be different radio access networks with their own radio resource control, but a common means of network access using a SIM will prevail. A top-level network architecture is shown in Figure 15.7.

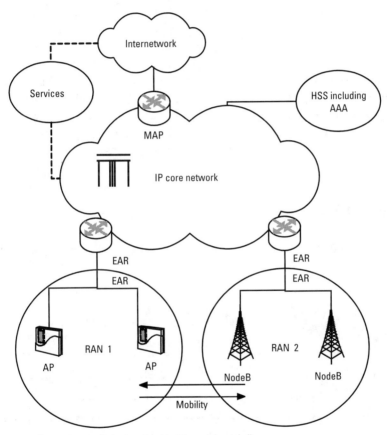

Operator controlled subscriber identity module or similar
AAA: authentication, authorization, and accounting
AP: access point
EAR: enhanced access router (i.e., AR with radio resource management and other capabilities)
HSS: home subscriber server
MAP: mobility anchor point
RAN: radio access network

Figure 15.7 A network architecture for B3G.

All this also makes one think of the value chain the future and the flow of money. One can easily expect that in the B3G-era different parts of Figure 15.7 will be owned by different stakeholders. For example, the AP could belong to the user or even to a radio access service provider in a building owned by someone else. The connection from the AP to the EAR could be that of a fixed operator. The service then could be provided by a separate service provider. Further complexity could be brought by the user being actually subscribed to a stakeholder than those already mentioned (i.e., another mobile network provider—a core network provider?).

A possible value chain and flow of money is shown in Figure 15.8. Each stakeholder will have different pricing, accounting, and payment methods; this all has to be taken care of and, in extreme cases, dynamically. This is already becoming a necessity as IMS is becoming reality in the form of next generation networks or fixed mobile convergence. Fixed rate can be applied, but then the question will be for which stakeholder of the value chain? Pricing could be in the form of what the airline industry uses—the best out there but opaque. The future of telecommunications could be to develop pricing or compensation strategy that can be as flexible but still transparent. Perhaps like hotels?

15.4 Technologies

The previous section gave a definition of B3G and what will come beyond that. Let us now have a deeper look at the definitions.

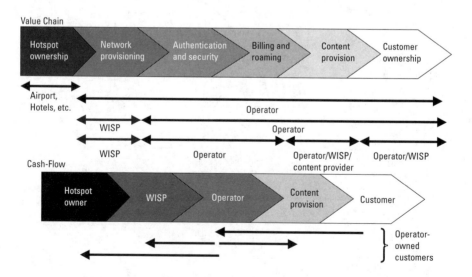

Figure 15.8 Possible value chain and cash flow.

15.4.1 B3G

B3G or interworking of different technologies is a thought born around 1998 when the first IEEE 802.11 products had just started shipping and i-mode was just picking up (or planning to).

Figure 15.9 shows the envisaged development in stakeholders of various networks and technological development for the short-, mid-, and long-term future. The figure also points out several technological issues that should be worked on. Arrows between two cells of the figure show the possibility of handover between the two technologies, while the shade of the arrow (gray scale) shows the expected extent of handover.

Integration of technology will provide adequate services to a user depending on mobility and availability. Of course this by itself brings along several new challenges (e.g., handover/handoff or mobility, security, and QoS). One of the main challenges of B3G is seamless handover, which should be provided while a user moves from the network of one access technology to the other and domain of one stakeholder to the other. Seamless handover means that the user does not perceive any disruption in service or quality even during handover. This topic, seamless handover, by itself brings in the study of several issues like security and QoS, which, in turn, should be done at each protocol layer and network element. The topic itself will require further study on development methods and technologies, including hardware, software, and firmware and technologies like application-specific integrated circuits (ASICs). Another important research

	Stakeholder -for handover -different parts (core, access, service)	IP	Broadcast (DVB, DAB etc.)	WWAN (3G, 2.5G etc.)	WLAN (IEEE 802.11)	FWA (802.16 etc.)	WPAN and Ad-hoc (IEEE 802 .15 etc.)
Short term (~3 yrs.)	One or more	v4 & v6	Similar to TV or radio and DVB-H	3G and 2.5G, handover: maybe	b,g,a,n,s, MAC enh.	WiMax mobile WiMax	
Mid term (3–5 yrs.)	More than one	v4&6	As above	3G →S3G and 2.5G, handover: possible	g,a,n,s,NG QoS, etc.	WiMax mobile WiMax NG	
Long term (5–10 yrs.)	Multiple	v4&6	SDR	2.5G, 3G, 3.5G and 4G handover: must	g,a,n,s, NG,QoS, etc.	WiMax mobile WiMax NG	UbiCom

NG(+)   Next Generation and beyond Handover
◀▬▶   For handover (arrows in 3 levels), it signifies the expected extent of handover.
Gray   The darker the arrow, the more common the handover between the concerned technologies.
    For technologies (e.g. 2.5G) it signifies lesser or decreased used of technology.

Figure 15.9 Envisaged technology development in short-, mid-, and long-term.

topic is software defined radio (SDR), which includes reconfigurability at every protocol layer [2]. Although we are discussing research, at least in Japan, a tangible/practically realizable and implementable SDR platform is being developed. Work on cognitive radio (e.g., activity of Defense Advanced Research Projects Agency Next Generation (DARPA XG) and 802.22) will also lead to further enhancement of work on SDR. In terms of standardization, IEEE 802.21 is working on the issue of handover for 802-based technologies, while system architecture evolution (SAE) work in 3GPP is doing the same. One should note that seamless handover is just one step; the goal is to achieve ubiquitous communication (UbiCom) [7].

WLANs provide handover within LANs, and work is going on toward further enhancement in this field. While WWANs provide handover, too, the challenge now is to provide seamless roaming from one system to another, from one location to another, and from one network provider to another. In terms of security, again both WLANs and WWANs have their own approach. The challenge is to provide the level of security required by the user while mobile from one system to another. Users must get end-to-end security independent of any system, service provider, or location. Security also incorporates user authentication that can be related to another important issue: billing. Both security and handover/mobility must be based on the kind of service a user is accessing. The required QoS must be maintained when a user moves from one system to another. Besides maintaining the QoS, it should be possible to know the kind of service that can be provided by a particular system, service provider, and location. Work on integration of the WLANs and the WPANs must also be done. The biggest technical challenge here will be the coexistence of the two devices, as both of them work in the same frequency band.

Fixed wireless access (FWA) is a technology that should be watched as it develops; depending on its market penetration and development of standards, it should also be integrated together with other technologies.

Figure 15.9 does not show the development of terminals and terminal technology, nor does it show development of satellite communications.

B3G will also see the convergence of computing and mobility together with networking. Terminals will be integrated with various technologies (e.g., computer and mobile integration with cameras). All this will require, as mentioned earlier, the field of ASICS, but at the same time the need for power will also increase; thus, development in the field of battery is a must. There is development in the field of low power and increased layers of circuit boards for miniaturization. The tasks to be performed will also require development in the field of CPUs and operating systems. More reliable production and cheaper displays that can fulfill the quality needs of multimedia services is a must too.

In terms of radio access technologies, all that was discussed previously basically means better use of the limited resource: the wireless spectrum. A couple of years back, the U.S.-based FCC launched the idea of cognitive radio (sometimes known as agile radio), which basically means use of any frequency band dynamically [18]. The point being that at a given time only 10 percent of all the available spectrum is used. This necessitates the development of SDR. Already various practically usable solutions and platforms for SDR are being developed globally.

15.4.2 Beyond

There are several technological developments that will take us toward the true era of UbiCom. In this section, such developments for the future are discussed. First, a definition of UbiCom is given.

15.4.2.1 UbiCom

Ubiquitous computing (ubicomp) is a term coined for a situation in which small computational devices are embedded into our everyday environment in a way that allows them to be operated seamlessly and transparently [7–10]. This means that many small objects/devices have the power of computing—everywhere. These devices are suggested to be active and aware of their surroundings so that they can react and emit information when needed—the devices can communicate with others. Ubicomp is not observable, though it is perceivable to human beings (i.e., it is unnecessary to make humans aware of its existence). However, interaction/communication between devices and humans should not be excluded.

UbiCom means that a human being or a device can do communication everywhere, anytime, whether it is a person or a mobile phone controlled by human or autonomous devices, through either wired or wireless links, satellites, or even ad hoc networks, or other means [7]. This is the description from an individual's perspective. Communication/interface between human users and systems (devices), if necessary, should be as natural as possible—like human conversation, supporting audio/visual and other natural means of information exchange. (Mobile) platforms can talk to other devices everywhere when necessary; communications and computing should also be context-aware (e.g., knowing the location and time). UbiCom requires interworking solutions that enable users to use and roam between heterogeneous (multiple types of) wireless networks.

Ubiquitous networking is one of the means to support ubicomp—communication networks exist everywhere.

Ubiquitous service could be used to describe the new services developed based on ubicomp. It could also be used to describe the services that are accessible everywhere at anytime.

15.4.2.2 Personal Networks

Another area of research for the next generation communications will be in the field of personal networks [11]. A personal network provides a virtual space to the users that spans over a variety of infrastructure technologies and ad hoc networks. In other words, personal networks provide a personal distributed environment where people interact with various companions, embedded or invisible computers not only in their vicinity but potentially anywhere. Figure 15.10 portrays the concept of personal networks. Several technical challenges arise with personal networks, besides interworking between different technologies, some of which are security, self-organization, service discovery, and resource discovery [11]. The European research project known as My Personal Adaptive Global NET (MAGNET) is working on a solution for personal networks [17].

15.4.2.3 Ad Hoc

Ad hoc wireless networks do not need any infrastructure. In these systems, mobile stations may act as a relay station in a multihop transmission environment from distant mobiles to BSs. Then they need infrastructure if they've got BSs. Mobile stations will have the ability to support BS functionality. The network organization will be based on interference measurements by all mobiles and base stations for automatic and dynamic network organization, according to the actual interference and channel assignment situation for channel allocation of new connections and link optimization. These systems will play a complementary role to extend coverage for low-power systems and for unlicensed applications. A central challenge in the design of ad hoc networks is the development of dynamic routing protocols that can efficiently find routes

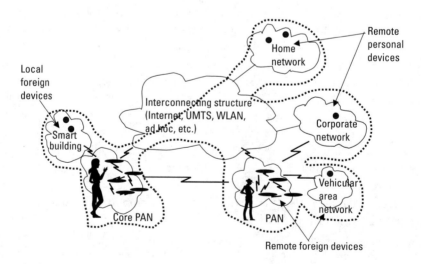

Figure 15.10 Personal network [11].

between two communication nodes. A mobile ad hoc networking WG has been formed within the IETF to develop a routing framework for IP-based protocols in ad hoc networks. Another challenge is the design of proper MAC protocols for multihop ad hoc networks. There are several other research activities going on in the field of ad hoc networks. IEEE 802.11s is working on solutions at the MAC layer. IEEE has named this technology *mesh technology.*

15.4.2.4 High-Altitude Platforms

High-altitude platforms (HAPs) have been proposed for a variety of applications ranging from communications to monitoring and sensing [12]. From a communications perspective, the relatively low altitude of these platforms (15–30 km) enables ultrahigh capacity communication to small ground-, air-, and sea-based terminals. Links to satellites are desirable for connectivity between metropolitan areas and islands of terrestrial cellular networks, as well as providing global area networks, where infrastructure is otherwise thin or lacking.

HAPs offer the potential for ultrahigh capacity, extremely high frequency HAP-ground links (to hundreds of megabits per second, depending on terminals), due to the low altitude of the platform. This enables high-capacity communication with extremely small, potentially mobile, terminals (e.g., consistent with handsets). Optical crosslinks and satellite uplinks are largely above atmospheric and rain attenuation, which would otherwise substantially degrade link performance and availability. Additionally, HAPs are well suited for providing full, high-capacity, multimedia information services over small, densely populated areas. Due to relatively low delay and delay variations, they can be more readily integrated with existing networks than satellite links, enabling reuse and optimization of use of existing infrastructure and technology.

15.5 Other Technologies

Several other developments are ongoing besides the development of mobile communications [13]. These developments will have an effect on next generation communications. In this section, a few of these technologies are described briefly.

One such technology is nanotechnology, a technology of building things small and atom by atom. This will affect the medical side of computing. There is also development in fuel cells (already available in the market) and fusion cells, which will make energy very efficient. Another development is the paper battery, which is very flexible and even foldable.

On the other hand, there is also development in the field of display technology—not only with foldable displays but work is ongoing toward holographic ones. There are also solutions existing for user interfaces where a keyboard, for example, is projected and the user can simply type on the projected image.

Meanwhile the development of holographic memory should allow much more memory space in smaller spaces than available today. Along the same lines, there is development in the field of 3D imaging.

Another field of development is quantum computing, communication, and cryptography. This technology hopes to deliver tremendous computing power. Quantum cryptography is already available.

When talking about small, we should also discuss sensor technology, including smart dust. The use of smart dust might be in different arenas; one of them is, for example, in cars, including the tires. Talking about smart technology also brings us to smart fluids, which can change shape depending on the electric charge applied to them. Changing shape while being strong certainly opens many doors.

Then there is the field of virtual reality, which is still active; research work is also going on in artificial intelligence (not mixing the two). This also brings us to robotics. Robots are already in industrial usage and are developing faster to bring solutions for the home market.

There is also tremendous growth in the field of genetics. Solutions will emerge for various diseases while research on genetics is ongoing; there are also developments in the biotechnology field focusing on internal organs and body parts.

A lot of work is going on in the field of complex systems in terms of analyzing and modeling. This research can help us to make better predictions. Also, let us not forget the developments in the field of materials concerning the provision of negative refractive index; this could affect the future of telecommunications as well.

15.6 Future Development: Protocol Layers and Technologies

In this section some technological developments required for future generation communications are discussed [1, 14–16, 19]. There are topics we have not covered, including MIMO-based technology and software and adaptive antenna technologies. The section is divided in protocols layers and component technologies.

15.6.1 Protocol Layers

Instead of discussing the OSI model, in this section the protocol layers are roughly discussed in the TCP/IP fashion. Instead of application layer, source coding is presented, and, together with physical layer, channel coding is also discussed.

In terms of source coding, the primary issue comes from the delivery of various services, some of them having very high capacity and demands, over a heterogeneity of networks, deployments, and architectures. For the case of source coding, flexible and scalable coding techniques are required with high compression and low complexity. The path toward a solution for such conflicting requirements could be to have the core source component (core of audio or video) of most perceptual significance to be coded aggressively and in an error-resilient way. Adaptability can be achieved in terms of coding rate, bit rate, other information to increase source quality and delay. Coding with consideration of communication protocols and network conditions is needed.

The TCP takes care of end-to-end performance. Originally TCP was designed for low error networks and, one could add, low congestion. Studies have shown that average error rate is an incomplete performance metric at packet level. It has strong dependence on second order statistics.

At the IP layer, the main issue is mobility as defined by MIP, it was not meant for fast or seamless handover. On the other hand, standardization bodies like 3GPP have defined IPv6 as the addressing scheme, but IPv6 has been barely implemented in most products. The other issue is resource reservation and QoS that is provided primarily by routing protocol; however, current routing protocols are not QoS agnostic.

Current MAC protocols are designed without consideration of their dependence on the physical channel characteristics. Most of all, the study is needed on power-efficient MAC protocols where today the power is mainly consumed during transmission. For proper design of a system, it is necessary that MAC protocol take in account not only the channel characteristics but also the physical layer methods like the antenna beam forming. Together with that, when considering the radio channel characteristics, it is necessary to look into various factors like bit error rate together with the signal-to-noise ratio information.

Channel coding adaptation based on carrier-to-signal interference information is required for improved performance. At the same time, adaptive modulation can further add up on the performance; this can be further enhanced when using hybrid automatic repeat request schemes. The gain of these solutions depends on channel conditions; thus, reliable prediction of channel conditions is needed.

Multicarrier (MC) techniques will play important roles in 4G systems; however, this means research on several topics. For instance, the optimum access protocol is unclear—OFDM-TDMA, OFDM-CSMA/CA, as well as MC-CDMA systems are all candidates. The performance comparison of these systems in multiple and isolated cell environments will be required. Adaptive array antennas can enhance the transmission performance for OFDM-based systems, but there are many different ways to configure on array antenna and OFDM demodulator. MIMO-OFDM has been the study of interest lately, but in open literature there is not capacity analysis of a MIMO-OFDM system that can jointly suppress cochannel interference from other cells. Besides the MC methods discussed in this section for high data rates, there is also work needed and ongoing in the field of UWB.

15.6.2 Component Technologies

Component technologies are those that form a system, likeradio resource management (RRM), QoS mechanisms, security, mobility, and routing. In this section, these technologies are briefly discussed. Together with the issues in each protocol layer, component technologies studies form a complete system.

Within RRM the call (or connection) admission control (CAC) supports access to a service at a given quality provided by a network. For a proper operation of CAC, it is crucial that the network can estimate as accurately as possible the consumption of the radio resources of a service and the current state of the wireless system in terms of traffic load, interference conditions, and capacity costs. The wireless network planners will have to design the appropriate thresholds for the CAC decision logic in order to have satisfactory radio resource utilization and at the same time the desired grade of service and coverage. Other important functions of the RRM are traffic scheduling, transport channel allocation and switching, handover control, and link adaptation. The tradeoff that characterizes the RRM is between satisfying the QoS requirement and optimal usage of the radio resources. Finally, the role of common RRM is foreseen as crucial in heterogeneous wireless networks. Here issues such as traffic addressing, handover control, allocation of wireless link over the most optimal radio access network based on capacity consumption, and the resulting cost for the end user represent interesting fields for further research.

As mentioned in the previous paragraph, there is the need to maximize wireless and device resources (battery, CPU, memory) usage while providing the best possible QoS. The biggest challenge is that each protocol layer tries to adjust the quality based on different measurements and by adjusting different parameters (e.g., TCP and MAC measure packet loss and adjust quality using automatic repeat request or packet size, while IP does the same using the route, and in physical layer it is the bit error rate or signal-to-noise ratio that leads to adjustment of the transmit power or data rate). The problem basically is to optimize resource usage for best possible QoS while maximizing channel usage and minimizing battery usage. This should be achievable by considering the adjustment of quality at different layers from application layer point of view; achieving the adjustment by knowledge of user-perceived quality will be the perfect solution.

Security is of utmost importance for the future generation systems to succeed. It is the core for a business to survive even today. When talking about B3G the first thing to understand is the relationship between the different administrative domains that are involved. There might be situation where such a relationship needs to by built on the fly. Added to it are the issues related to authentication, different levels of authorization, key management while mobile, accounting and charging, fraud management, privacy issues, security of the network infrastructure, and of course secure attachment and detachment from a network, together with secure configuration of the device. On top of these are the standard requirements on security of lawful interception, scalability, and management. Then there are requirements like maintaining the security level during mobility; impact on the network and impact of the security solution on the resources of the network and the device must also be considered. When talking about seamless handover, all the steps related to security should be fast while still providing the required level of security. Issues related to seamless handover are discussed in Section 15.9 [20].

Ad hoc networks are expected to play an important role in the future. There are several security issues that the ad hoc networks will face; they also face issues related to the variety of devices that can communicate with each other; some of them might have a powerful CPU or a lot of memory, while others might exist that have severely restricted (peanut) CPU and battery power.

Routing has been touched upon in an earlier section already. The main issue here is to provision a QoS enabled routing.

Mobility will be of the utmost importance in the future. Here, mobility comes with seamlessness. The user should not perceive any change in service quality when performing handover from one network to another or from one technology to another. What we thus see is that mobility is intertwined with QoS and security, and these three cannot be separated. This brings in the need for a study of the complete system instead of one component (mobility, QoS, security, or routing).

15.7 Wireless Standards Activities Toward the Future

Today basically three wireless technologies, besides satellite communications, have made an impact: WLANs, WPANs, and WWANs. WMANs, on the other hand, have created a hype recently, which it seems will grow and might make even bigger impact. WLANs complement LANs, while WPANs are used for short distance communications, and WWANs cover wide areas and are most commonly known as mobile or cellular communications. WMANs are meant to give coverage like WLANs but outdoors, although it seems from recent standards and activities that this technology can become competition for both WLANs and

WWANs. Besides that, recently WLANs are being seen as a threat to the WWANs but in fact these two are complementary technologies. In the following, the future direction of WPANs, WLANs, WMANs, and WWANs are presented; an overview of wireless technology standards is given in Table 15.1.

15.7.1 WPANs

Besides the WLANs, the WPANs like Bluetooth, HIPERPAN, and IEEE 802.15 are standardized. These technologies will be used for short distance ($\sim$10m) communications with low data rates for different QoS. It is envisaged that the WPANs will exist in all the mobile terminals in the near future. The WPAN standards, IEEE 802.15.3 and .3a, have developed, and work is ongoing on higher data rates of about 55 Mbps, thus paving the path toward broadband WPANs. IEEE 802.15.4 is focusing on very low data rate solutions, which will work at a few or a few hundred kilobits per second, which is a first step toward the development of body area networks. Several companies have reached a consensus on UWB as a low-data-rate solution for IEEE 802.15.

15.7.2 WLANs

Today LANs mostly access the Internet using IP. The growth in wireless and the benefits it provides has brought forward changes in the world of LANs in recent

Table 15.1

Wireless Technologies

WWAN	WMAN	WLAN	WPAN	Cordless
GSM-HCSD, GPRS, EDGE (WAP)	IEEE 802.16, IEEE 802.20	IEEE 802.11	IEEE 802.15	PHS
IS-95	*HIPERACCESS*	*HIPERLAN/2*	Bluetooth	DECT
IS-54/IS-136	*High-speed wireless access*	*MMAC Ethernet WG and ATM WG (HiSWAN)*	HIPERPAN	CT2/CT2+
PDC (i-mode)	*BWIF*	MBS		
3GPP, 3GPP2 (HSDPA, HSUPA, IMS)	LMDS	MMAC wireless homelink		
3GPP (LTE and SAE)	MMDS	HomeRF 1.0 & 2.0		

Legend: Text in italics means technology does not have much market penetration or is dead.

years. WLANs provide much higher data rates than WWANs for slow mobile or static systems. The IEEE 802.11b-based WLANs are already being widely used, while the IEEE 802.11g and IEEE 802.11a are also available.

WLAN technologies are mainly used for wireless transmission of IP packets. Until now, in contrast to the WWANs, the WLANs provided network access as a complement to the wireline LANs. In the near future, QoS-based WLANs are expected to come onto the market in the form of IEEE 802.11e. Security in IEEE 802.11i is already available, although work on management frame security is still ongoing. The IEEE 802.11 WG has also accepted a mobility solution known as inter access point protocol, IEEE 802.11f. Another group in IEEE 802.11 is working on RRM (IEEE 802.11j). The IEEE 802.11 committee has approved IEEE 802.11h, dynamic frequency assignment, and transmit power control. Due to the success of the standard, several other study groups are looking at higher data rate solutions (IEEE 802.11n 110 Mbps+) and next generation technologies, including standardization work with 3G standardization committees.

WiFi Alliance, an industry alliance, is providing interoperability specifications and tests of the IEEE 802.11 products for better acceptance in the market. This alliance also provides recommendations for roaming between different WISPs so that a user, or customer of one WISP, can access WLAN services when in another WISP's hotspot and still receive one bill.

Harmonization in 5-GHz-band technologies is a must so as to avoid making the 5-GHz band a garbage band. For the time being, the success of a standard will depend on pricing, performance, availability, and marketing of the standards.

Besides the work being done by the standardization committees, there should be a study on providing top-to-bottom mapping. The correct mapping of higher layer protocols to lower layer protocols is a must to provide optimum service. Especially in the case of IEEE 802.11, where the standard only defines the bottom two layers, relations must be created with IETF, the committee developing layer three and some higher layer protocols.

Basically, most of the current development will lead to providing users different services within WLANs; in other words, it is the integration of services within one system. Another step currently becoming visible is toward integration with WWAN technologies like 3G and WMANs like mobile WiMAX.

15.7.3 WMANs

WMANs have long existed, but only recently has this technology started seeing true success—the technology is also known as fixed wireless access (FWA) or broadband wireless access (BWA). Failure in past came mainly due to nonstandard solutions. Today, the standard is ready in the for of IEEE 802.16.

The more recent version of 802.16, IEEE 802.16e, is much secure and provides mobility at high speed. Thus WMAN technologies can provide high data rates while the user is mobile.

Similar to Wi-Fi, the IEEE 802.16 community has started WiMAX for interoperability and testing purposes. WiMAX has also adopted mobile WiMAX for IEEE 802.16e.

Within IEEE 802 there is the activity of mobile broadband wireless access (MBWA) in the form of IEEE 802.20. This group is developing solutions that start from a clean slate without the need for backward compatibility.

15.7.4 WWANs

Growth in the field of WWANs, more commonly known as mobile communications, has been tremendous in the past decade. Second generation (2G), 2.5G, and 3G standards of mobile systems are being used, while efforts are going on toward development and standardization of B3G systems. The existing (2G) systems are mainly used for voice purposes. Due to the tremendous growth of the Internet, some support for data services like WAP and i-mode has been developed. 2G supplement systems, or 2.5G, like GPRS and now 3G systems, provide further possibilities for data services with varying QoS requirements.

At present the main application for data services over mobile communications systems is Internet access. The future is toward a full multimedia-type application providing various levels of QoS using an IP-based backbone. Thus, WWAN is also moving toward integration of services.

Further work is being done by the standardization committees to integrate other access technologies with 3G. Another development in the standardization of WWAN is toward an IP network—3GPP SAE activity—and higher data rate in the radio access network (RAN)—3GPP long-term evolution (LTE) activity. All this shows us that the WWANs are moving toward packet-switched solutions and integration of technologies now that integration of services has almost been achieved.

15.8 Standardization and Regulations

As future generation communications work is still in its infancy—the standardization work is not yet happening. Still one can see different things.

IEEE 802.21 is working toward B3G and is the only true standardization work in this field. 3GPP and 3GPP2 are also looking into 3G and WLAN integration and in a way can be seen as standardization work for B3G. It remains a question whether 3G standardization activities will become the base for 4G standardization. Although 3GPP activities in the future toward all-IP and seamless handover between 3G and WLAN does make one think so.

Now different countries or regions have started work on 4G, like India, China, Korea, Japan, and Europe. In fact China, Japan, and Korea have formed a cooperative to develop next generation solutions. In Japan a mobile IT forum has formed to work toward 4G. In Europe, WWRF is working on 4G and so are the European 6th Frame Work Projects. WWRF is now becoming a worldwide forum. FUTURE in China is doing similar work. A next generation mobile communications group formed to discuss 4G and B3G especially in China, Japan, and Korea [16].

The FCC has provided the 255-MHz unlicensed band in 5 GHz, and there is a proposal for the 3.65-GHz band to become an unlicensed band. There are also thoughts to make licensed bands available at 70, 80, and 90 GHz.

ITU-R has already discussed further development of IMT-2000 to 30 Mbps. There is also talk of 100 Mbps for high-mobility and 1 Gbps for low-mobility systems. Further the work on interworking between different air interfaces remains active.

A more recent activity related to next generation networks is looking into convergence of fixed and mobile networks. The goal here is to use IMS for service while different technologies provided by fixed and mobile operators can be used to access the service. Access of service through different operators and from another operator also has to be provided while performing handover. There are different next generation network activities; one of them is the ETSI TISPAN group, which recently finished its first set of specifications.

3GPP is currently working toward much higher data-rate solutions known as super 3G, and thus there is ongoing work toward a radio access network while the core network is being changed to all-IP. Both these enhancement activities are also known as SAE and LTE.

15.9 Mobile Networks Security Issues

In this section we will look at security issues related to mobile networks. The focus in this section is on security issues related to seamless handover in B3G.

15.9.1 Introduction

In Figure 15.11 the common steps related to mobility are given together with the associated delays and possible security issues. For a given radio access technology or network, not all security issues will exist, and some delays might be negligible.

For the study of secure and seamless mobility, we consider the issues listed next. The last three are of lower priority for us; thus, they are not discussed further in this section. (See Figure 15.12.)

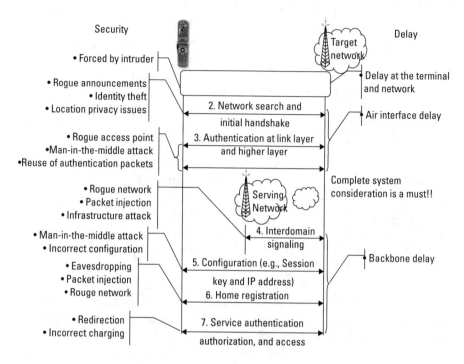

Figure 15.11 Generic mobility steps, delays, and security issues.

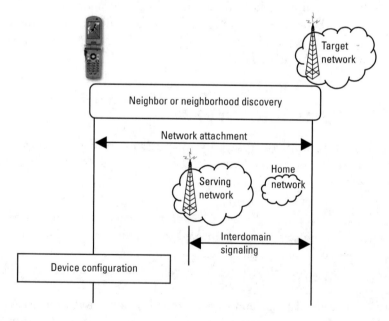

Figure 15.12 Security problems mapped to the message sequence.

- *Network attachment:* network search, association, and authentication;

- *Neighbor discovery:* layer 2 or layer 3, designed to be within a technology;

- *Neighborhood discovery:* different technology or operator network;

- *Device configuration:* IP address and session key;

- *Interdomain signaling:* communication between networks, including dynamic agreements.

For the complete system to work, it is necessary to see how these solutions fit together and their effect on other components like QoS and routing. Further, secure solutions should be developed for handover decision algorithms.

15.9.2 Neighbor Discovery

Neighbor discovery is performed when a device connects to the network, where the connection could be that of a mobile due to handover or connection of a new device directly to the network. The new device connecting directly to the network could be, for example, an AP. The AP would need to find other network elements like the switch or router and possibly other APs too. Thus the depth of neighbor discovery in terms of number of hops also needs to be defined. In the case of handover, the steps related to neighbor discovery are

1. Finding that the service quality has reached a level in which a trigger for neighbor discovery is needed;
2. After getting the trigger, to perform active, by sending probes, or passive, listening to beacons or other messages, search for neighbors.

When discussing mobility or handover, neighbor discovery is the method to find the network element to which an ongoing communication or connection can be moved. The reason for this movement could be due to mobile platform movement or any other reason leading to such decision either in the network or the mobile device (e.g., lost signal). For layer 2, it means discovering an AP or a bridge, while in layer 3, this means discovering a new router or other network element in the neighborhood (e.g., an AP discovering a router). Discovery of a rogue AP and handover to it could, for example, lead to man-in-the-middle attack leading to revenue loss for the operator or an increased bill for the subscriber. If the neighbor discovery process takes a long time, the user will experience service quality degradation or dropped connection, leading in turn to lost revenue and in the worst-case scenario, increased churn rate. Secure, fast neighbor discovery is a must for achieving secure and seamless handover.

Solutions for layer 2 (IEEE 802.1ab, IEEE 802.11f, and SNMP) and layer-3 (SEND) are available. There are at least a few issues with these solutions:

1. None of the security analyses of these solutions are available, especially for the case of MNO and hotspot deployments.
2. The study on delay associated with these methods is not available.
3. The use of these solutions for seamless mobility is not considered (i.e., how do these solutions fit in the complete picture?) This also means that optimization work needs to be done.
4. The available solutions for layer 2 and layer 3 do not give one set of solutions, and there is no study showing whether they can work together for either network-controlled or mobile-controlled handover.

15.9.3 Neighborhood Discovery

Neighborhood discovery is a method to find radio access technologies and domains other than the serving one to which an ongoing communication or connection can be moved (i.e., for intertechnology or interdomain handover). For initial study, the technology in the neighborhood could be either IEEE 802.11 or 3GPP. As with neighbor discovery, neighborhood discovery requires solutions for both layer 2 and layer 3, but higher layer discovery in terms of checking the relation between two domains will also be necessary. The security and delay issues are very similar to that for neighbor discovery, and obviously a fast and secure solution is a necessity. The steps for neighborhood discovery are same as those for neighbor discovery.

IEEE 802.11u, 3GPP, and 3GPP2 are looking into solutions for intertechnology handover, while IEEE 802.21 is also developing solutions for interdomain handover. IETF work on neighborhood discovery, media-independent handover enabling protocol, and Seamoby have started or exist. Still none of the solutions are available today, and it is unclear when the work will be completed or even whether the solution is good enough.

15.9.4 Network Attachment

Network attachment procedure relates to both the initial network access and network access after the handover. For both cases, at least part of the neighbor or neighborhood discovery must be initially performed.

Attachment to a network involves several steps, such as association, authentication, access control and authorization, home registration, assigning of IP addresses, and communication at the service provider level. During the attachment phase, several parameters are communicated and decided on (e.g., authentication type and available data rate). The steps of network attachment lead to several security attacks, like DoS, man-in-the-middle, unauthorized access of service, and lack of privacy solutions. There is also the issue of identification associated to these attacks. Solutions for each step involved in network attachment exists separately, as does the security issue, and considerable delay is associated with the steps.

15.9.5 Device Configuration

Configuration of the device means setting up the IP address and the session key for confidentiality and integrity. Such configuration would be required for each session or each time the mobile platform is attached to the network. Some configurations do not change very often and might be required once in the lifetime of a device, such as setting the SSID of the network to be accessed (could be given during network attachment), username, and password. Insecure procedures could lead to incorrect configuration. This means that incorrect IP addresses or session keys could be configured, leading to a DoS attack. Further, the configuration process could incur long delay.

IETF has a Zeroconf solution, and there are other solutions that make use of near field communication to transfer the required information to a device, and Microsoft has a wireless configuration solution. None of these solutions are meant for operator-owned network.

15.9.6 Interdomain Signaling

Interdomain signaling is required if a device is in or moves to a foreign network or if service is being accessed from a foreign network. This could include communication via an intermediary network. When handover occurs, normally the target network will communicate with the serving network, or more appropriately the home network. This communication or signaling can also go through a broker in case there is no contract between the two domains.

Signaling in the backbone and between different domains certainly means long roundtrip delays. If this involves a broker, then the overall delay can add up to several seconds of course, depending on the network and network conditions. One could also face a man-in-the-middle attack or packets could be injected in the network.

Broker-based solutions or direct relations between networks exist, but this is only static solution and not the right way to the future, where connection should be more dynamic in nature especially due to the large number of networks.

15.10 Mobile Platforms Security Issues

While mobile platforms are proliferating and are increasingly becoming important for traveling business people, companies, hospitals, and a variety of other agencies and institutions, these devices are becoming repositories for more and more sensitive data and/or high-value content. However, because of their size, these mobile platforms are much more vulnerable to loss or stealing. Therefore, protecting data and the content they carry is vital.

In order to reach this goal at a time when networks are converging; mobile platforms are becoming more processing-capable, more communications-capable, and more open; mobile platforms support more over-the-air provisioning; and mobile platforms need more software updates and reconfiguration, it is necessary to address security from end to end.

First of all, it is important to control the network access and prevent unauthorized access in order to protect the service provider's revenue, protect the network properties, and guarantee QoS to the user. Then, it is important to seamlessly protect the information flow to the wireless platform on any network. Finally, the mobile platform itself needs to have enough trust to guarantee the integrity of the protection of the user's data and applications and the service provider's applications. As described earlier, this trust cannot take place unless the platform has a secure execution environment, protected storage, validated system software, authenticated applications, and authorized user transactions. These requirements will remain unchanged in future mobile platforms; albeit how they're implemented will improve.

Furthermore, there are new and emerging threats. Malicious software is getting better at launching DoS attacks on network entities and damaging devices. Trojan horses' ability to destroy mobile platforms by modifying its applications is gaining popularity. Also, while users becoming more privacy-aware, intrusions can compromise user privacy. With more valuable content and more sensitive data placed on the mobile platform, such platforms will naturally morph into an attractive target for hackers, cyberterrorists, and crime organizations. These attacks will imply revenue loss and user-privacy intrusion.

Finally, protection against those threats of tampering and unauthorized access needs to happen in very complex environment: increased legal liability, government legislation, and privacy regulations and guidelines at the state, country, and international levels.

So what security capabilities are needed for future mobile platforms? The answer is this: the necessary technologies and controls to guarantee the necessary levels of protection for the data and services enabled on those platforms. Through tamper-resistant safeguards, sufficient application isolation, intrusion detection and prevention solutions, and protected access control mechanisms, mobile platform will be able to protect data from potential malware. As these mobile platforms become more ubiquitous, they will also require distributed trust management without having to rely on a central authority for trust decisions.

15.11 Conclusions

This chapter discussed the next generation communications technology. Definitions of 4G, B3G, and future generation communications were given in the chapter. The technology for the future should be acceptable to the users and sellable by vendors and operators. It is also important to understand the various technologies that are being developed in other fields, as they also affect the usage of telecommunications-related products; these too were briefly discussed. The chapter also discussed various technical challenges and problems to be solved for next generation communications.

One of the common behaviors in any research or development is the thought of components or layers instead of a complete system. A developer of an application layer, for example, thinks that all lower layers will work perfectly fine. The fact is that there is a severe need for understanding and working from a system perspective. Looking at components is of course beneficial but only at the very starting stages of research. This stage can be called the organic stage, when thoughts are still developing and should be given space to develop without hindrance. Once this stage is past, which should not take long, boundaries should be set and reality should be brought in to play. Here the reality is that each component has to work in a system, and the boundary consists of the technological limitations and the system-level view. Many of the issues that are to be tackled for future generation communications can be solved by good communication between human beings, complete system views, and hard work—it does not require "rocket science." The most important point is of course that security considerations should be there from the beginning.

The chapter has discussed a lot about the future generation of communications, but most of it is, at least for some time in the beginning, of benefit only for the developed nations. It is extremely important to remember the developing nations, where a big part of the world's population lives. For developing countries, too, many similar services as in developed nations will be needed, but at a fraction of the cost. An average person in such a country is not able to pay more than US$5 per month for phone costs—anything more is impossible. This brings a whole paradigm shift in the research, design, and development of new products and thus the future generation of communications and their security concerns.

At the end, it is worth noting again that the main thing that will sell is person-to-person communication. Today peer-to-peer is a hot topic, and it is hot in terms of usage because it provides true person-to-person communication. After this is the machine-to-machine communication.

References

[1] Prasad, A. R., "WLANs: Protocols, Security and Deployment," Ph.D. Thesis, Delft, the Netherlands: Delft University Press, December 2003.

[2] Lauridsen, O. M., and A. R. Prasad, "User Needs for Services in UMTS," *International Journal on Wireless Personal Communications,* Vol. 22, No. 2, August 2002, pp. 187–197.

[3] Christensen, C. M., *The Innovator's Dilemma,* New York: Harper Collins, January 2003.

[4] Innovators, http://www.fortune.com/fortune/fsb/specials/innovators/cook.html.

[5] Quintillion Technologies, http://www.quintillion.co.jp.

[6] Gombrich, E. H., *The Story of Art,* Boston, MA: Phaidon Press, 1995.

[7] Prasad, A.R., P. Schoo, and H. Wang, "An Evolutionary Approach Towards Ubiquitous Communications: A Security Perspective," *SAINT 2004,* Tokyo, Japan, January 26–30, 2004.

[8] Weiser, M., "The Computer of the Twenty-First Century," *Scientific American,* Vol. 265, No. 3, September 1991, pp. 94–104.

[9] Weiser, M., "Hot Topics—Ubiquitous Computing," *Computer,* Vol. 26, No. 10, October 1993, pp. 71–72.

[10] WWRF, "The Book of Visions 2001, Visions of the Wireless World," v1.1b, December 2000.

[11] Niemegeers, I. G., and S. M. Heemstra de Groot. "Research Issues in Ad-Hoc Distributed Personal Networks," *International Journal on Wireless Personal Communications,* Special Issue on Wireless Personal Communication, Vol. 26, No. 3, 2–3, pp. 149–167, September 2003.

[12] Farserotu, J., et al., "Scalable, Hybrid Optical-RF Wireless Communication System for Broadband and Multimedia Service to Fixed and Mobile Users," *International Journal on Wireless Personal Communications,* January 2003, Vol. 24, No. 2, pp. 327–339.

[13] "Technology Quarterly," *The Economist,* March 11, 2004, June 10, 2004, and September 4, 2003.

[14] Prasad, R., and M. Ruggieri, *Technology Trends in Wireless Communications,* Norwood, MA: Artech House, 2003.

[15] Hara, S., and R. Prasad, *Multicarrier Techniques for 4G Mobile Communications,* Norwood, MA: Artech House, 2003.

[16] International Conference on Beyond 3G Mobile Communications-2004, Tokyo, Japan, May 26–27, 2004.

[17] IST MAGNET, http://www.telecom.ece.ntua.gr/magnet.

[18] FCC Cognitive Radio Technologies Proceedings, http://www.fcc.gov/oet/cognitiveradio.

[19] Prasad, A. R., and N. R. Prasad, *802.11 WLANs and IP Networking: Security, QoS, and Mobility,* Norwood, MA: Artech House, 2005.

[20] Prasad, A. R., A. Zugenmaier, and P. Schoo, "Next Generation Communications and Secure Seamless Mobility," *SecQoS 2005,* Athens, Greece, September 2005.

[21] A. R. Prasad, "The Future Re-Visited," Invited paper, *Springer International Journal on Personal Wireless Communication,* July 2006.

16

Mobile Security Threat Catalog

Security is about building protection into the device starting from the very early phases of its conception. Security should be a major goal in specifications and design; from there it should be taken into account in the development phase and finally tested for.

This chapter provides a list of major security threats and should be used as a reference to other sections in the book.

16.1 Software Threats

Malware [1] stands for malicious software and comprehends programs developed to do harm, as listed in Table 16.1.

Table 16.1

Software Threats

ID	Threat	Description
1	Backdoors	A hidden entrance to a mobile platform that can be used to bypass the system's security policies.
2	Buffer overflow	A method of overloading a predefined amount of space in a buffer, which can potentially overwrite and corrupt memory in the data and hence allow an attacker to insert his own program code into vulnerable software (including OSs) causing malicious damage. Buffer overflows can be stack-based or heap-based.

Table 16.1 (continued)

ID	Threat	Description
3	Password attack	An attacker attempts either password guessing or password cracking in order to obtain passwords stored on the mobile platform.
4	Rootkits	Memory-based rootkits install in active memory, so flushing memory or power-cycling a device renders the rootkit useless. Their potential useful life is short. Persistent rootkits become active each time a mobile platform boots. They install in the Windows registry or as part of the Windows file system. In general, this type of rootkit is associated with malware that initiates a specific action, like sending personal data to a remote location and continues to perform the operation until removed from the system.User-mode rootkits intercept data at the user level to avoid detection. When an application running as the current logged on user attempts to locate information, like the contents in memory, a user-mode rootkit attempts to disguise its existence by excluding itself from the results.
5	Spyware	A program that monitors user's actions and transfers information to third parties.
6	Trojan and bots/botnets	A malevolent piece of code hidden in a program that performs harmless tasks. (See Chapter 5.) Trojans can create bots.
7	Virus	A program that must be installed by user action and that infects other programs or data. A virus is self-replicating, malicious code that attaches itself to an application program or other executable system component and leaves no obvious signs of its presence. (See Chapter 5.)
8	Web application attack	An attacker attempts to perform account-harvesting, session tracking, variable alteration, or browser flaw exploitation to launch an attack on the mobile platform.
9	Worm	A program that is transmitted automatically by replicating itself and infecting other programs or data.

Published attacks on mobile platforms are reported in Table 16.2.

Laptops, PDAs, and smartphones used in standalone mode are not vulnerable to direct attacks, as malware spreads from the computer the device is connected to and/or the network to the handheld device or vice versa. Nevertheless, one of the great advantages of handheld devices is their ability to synchronize to a desktop computer and to connect wirelessly to the Internet so that a user can access important data on the go.

In Table 16.3 we will describe the synchronization operation for the two leading handheld operating systems, PalmOS and PocketPC.

Internet connection can be gained via a hardwired or a wireless network interface card. When a wireless connection is established, the security of the wireless means supporting the connection must be taken into account. Over-the-air attacks are more frequent than attacks through a wired connection, as

Table 16.2
Published Attacks

ID	Published Threats	Description
1	Trojan	Trojan horses can spread on mobile devices supporting Java architectures. Trojan horses can be hidden inside Java applets that a user may want to install on her mobile device at post issuance. Propagation of a Trojan horse on a GSM phone was described in [3]. Trojan horses can perform different types of actions, from disturbance through DoS to identity theft through disclosure of credential or personal information. The Trojan horse in [3] is hidden in a game for mobile phones, while the real goal of the program is to retrieve the SIM card PIN. The game can supposedly be downloaded onto the mobile phone and installed as Java application. When it is stored for the first time, the Trojan horse is activated and simulates a phone reboot, prompting the user to enter his PIN code. Since users of electronic devices are used to reboots they will most probably not suspect the action of malware and insert their PIN. Supposedly, backdoor code may transmit the PIN value to a malicious user. The victim will not be alerted, as after the first reboot the Trojan horse will not be active and the game will perform as expected. This is an example developed and described in [3] to show that Trojan horses can affect mobile phones, but other actions beside PIN disclosure could also have been implemented.
2	Worm (Cabir)	The first worm for smart phone, called Cabir, targeted Symbian terminals that support the Series 60 platform and was propagated through Bluetooth [2]. It was simply a proof-of-concept worm, with no harmful action. Cabir replicates through Bluetooth connections to phone messaging inboxes as the file caribe.sis. When the user installs caribe.sis, the worm activates, and the consequent action is that it will try to spread to new devices over Bluetooth. Cabir cannot automatically reach the target device; the victim must accept the file transfer while the infected device is in range. Cabir can only reach mobile phones that support Bluetooth and are configured to discoverable mode. Cabir is a slow-spreading worm; it will infect at most one device per activation, because once it finds a device it locks to it and won't look for others.
3	Worm (Lasco)	The new version of Cabir is called Lasco and is more harmful. Once installed, Lasco constantly searches for other Bluetooth devices, quickly draining mobile phone batteries. Lasco is capable of spreading in two ways: through Bluetooth like Cabir and by embedding itself into the SIS installation files and propagating when programs such as games are shared. Lasco's twofold propgation is typical with PC malware, but it is used for the first time in mobile devices.

Table 16.3

Synchronization Vulnerabilities

ID	Synchronization Feature (OS Type)	Vulnerability
1	ActiveSync (PocketPC)	The synchronization operation doesn't prevent malicious software exchange between the devices. The TCP and UDP ports opened to allow the synchronization operation may be exploited by malicious users to establish a connection to the PC. ActiveSync is password protected, and sniffing and brute force attacks to password authentication apply. (By default the number of password attempts is unlimited.)
2	HotSync (PalmOS)	The synchronization operation doesn't prevent malicious software exchange between the devices. The TCP and UDP ports opened to allow the synchronization operation may be exploited by malicious users to establish a connection to the PC and through it to the network it is connected to, often a corporate private network. Through these ports, confidential data may be accessed or malicious code propagated.

physical access to the device is not necessary. Table 16.4 reports the vulnerabilities of wireless means.

16.2 Hardware Threats

Table 16.5 lists several hardware threats.

16.3 Network Threats

Security of the network is dependent on the design of the network and the network and internetwork protocols that are being used, including signaling protocols in the case of switched network technology and routing protocols in the case of a router network. In brief, the threats for a network could arise from packet data corruption, packets delivered out of sequence, packet loss, packet duplication, misrouting, address spoofing, and packet replay/insertion. In Table 16.6 many of these issues are given in different forms of possible attacks for a network [4].

Table 16.4

Vulnerabilities of Wireless Means

ID	Wireless Means	Security
1	Bluetooth	Bluetooth may provide authentication and data encryption, but Bluetooth security is known to be very weak. Its main vulnerability is the dependence of all key calculations on a PIN that must be shared between the two paired devices. As users often use 4-bit PINs, all keys may be quickly derived by performing an offline attack and testing all possible PIN values. (See Chapter 11.)
2	Cellular	The communication security varies depending on whether the communication is GSM or CDMA, or 3GPP. GSM and CDMA provide user authentication and data encryption on the wireless link between the user device and the operator's BS. 3GPP provides user authentication, data encryption, and integrity for access security as well as for network security. (See Chapter 14.)
3	Wi-Fi	Wi-Fi may provide authentication, data encryption, and integrity if WPA-compliant devices are deployed and security mechanisms enabled. If devices do not support WPA, WEP should be enabled. It has been proven that WEP has many weaknesses, but it can still provide minimum security against casual eavesdroppers. By default, security mechanisms are disabled, so users should take care of enabling them before use. (See Chapter 12.)

Table 16.5

Hardware Threats

ID	Threat	Description
1	BIOS-modification attack	In this attack, an adversary attempts to modify the BIOS to execute malicious code such as inserted Trojans. (See Chapter 4.)
2	DEMA/SEMA attack	See electromagnetic analysis attack.
3	DMA attack	In this attack, an adversary attempts to repurpose built-in bus-mastering hardware to be not successfully executed. (See Chapter 4.)
4	Electromagnetic analysis attack (also called DEMA/SEMA attack)	In this attack, an adversary attempts to measure the electromagnetic radiation to exploit local information, and, although more noisy, the measurements may be performed from a distance.

Table 16.5 (continued)

ID	Threat	Description
5	Fault attack (also called glitch attack)	In this attack, an adversary attempts to induce a fault by some physical perturbation or by pushing a chip out of its valid operating conditions in order to cause an abnormal behavior of the chip or to make processing errors and then set these errors to find the cryptographic keys. (See Chapter 4.)
6	Glitch attack	See fault attack.
7	Identity-spoofing attack	In this attack, an adversary attempts to spoof the identity of the phone, and, if modified, the stolen mobile device may be given an IMEI replacement. (See Chapter 4.)
8	Invasive attack (also called physical attack)	In this attack, an adversary attempts to irreversibly modify the physical properties of a chip in order to capture information stored in memory areas or flowing over the data bus. (See Chapter 4.)
9	JTAG/SCAN attack	In this attack, an adversary attempts to use a chip's or a board's self-test capabilities, which are intended for debugging and post manufacture testing, to extract keys and other data by scanning and searching.
10	Key-spoofing attack	In this attack, an adversary attempts to spoof the private keys used by applications to authenticate themselves or to digitally sign data. (See Chapter 4.)
11	Memory-copy attack	In this attack, an adversary attempts to use a malicious checksum function in order to compute a checksum over a correct copy of a checksum function.
12	Memory-reprogramming attack	If an adversary has physical access to the mobile device, he will attempt to reprogram the memory. (See Chapter 4.)
13	Observation/modification attacks	In this attack, an adversary attempts to observe or modify user data (Chapter 4).
14	Physical attack	See invasive attack.
15	Power analysis attack	In this attack, an adversary attempts to study the power consumption of a cryptographic hardware device such as a smart card and to yield information about what the device is doing or even some key material. (See Chapter 4.)
16	Side-channel attack	Side-channel attacks include timing analysis attacks, power analysis attacks, and electromagnetic analysis attacks. See details under those threats. (See Chapter 4.)
17	Timing analysis attack	In this attack, an adversary attempts to infer information on the data or secret values due to a dependency between code execution time and data being processed. The adversary may also watch for the length of time a cryptographic algorithm requires or watch data movement into and out of the CPU or memory on the hardware running the cryptosystem or algorithm. (See Chapter 4.)

Table 16.6
Network Threats

ID	Threat	Description
1	Access control and authorization granularity issue	This is a problem where authorization is equated to authentication. A given user thus authenticates to a mobile network, and instead of only being allowed to get access to, for example, only voice, the user gets access to all other services in the network. In the initial 802.16 (see Chapter 13), authorization-related issues existed due to the problem with integrity and replay protection. Bluetooth devices have serious issues with authorization and access control in the form of Bluejacking or Bluebugging (see Chapter 11). This can also lead to DoS attacks. TLS also faces this problem, due to the weak integrity protection it employs. (See Chapter 7.) The original security solution of IEEE 802.11 (see Chapter 12) also had issues with integrity protection and access control, leading to authorization failures.
2	Bandwidth consumption attack	This is a type of distributed DoS attack. An attacker making use of multiple machines floods the ports with spurious TCP packets in order to utilize all available bandwidth in the network. Such flooding attacks require several machines to attack the router interfaces, causing legitimate packets to be dropped or discarded. (See Chapter 7.)
3	Elevation of privileges	This attack can be used in organizations but should be viable in IP-based mobile networks. This attack also relates to the authorization issue. The basic idea is that an attacker gets into a network and then, using various methods, tries to get the privileges of a user or network administrator with the most privileges (e.g., administrator). The methods for this attack in a mobile network could include, among others, bugs in network or software being used, network elements with manufacturer password for management unchanged, and the interfaces of network elements being available to users.
4	Internet control message protocol (ICMP) attack	The purpose of ICMP is for testing and providing error messages in an IP network, but as security consideration of the complete system is a must, ICMP has been exploited for several attacks. Some of the attacks are ICMP attacks causing DoS. ICMP sweep or ping sweep, on the other hand, is used to determine the hosts that are active on a network. There are several other attacks possible, like oversized ICMP packets causing crashes of target hosts, message redirects causing man-in-the-middle attacks, router discovery messages spoofing routers and hijack traffic, and ICMP tunneling leaking information from a system by setting covert channels.
5	Identity theft	This can be a big issue, as the mobile systems are embracing open platforms and off-the-shelf technologies, all leading to several identities. Binding of these identities correctly and enforcing this binding in correct fashion is a must. Not doing so will lead to several attacks like man-in-the-middle and DoS, among others. (See Chapter 2.)

Table 16.6 (continued)

ID	Threat	Description
6	Impersonation	A method to impersonate other users is known as impersonation. This way, obviously, the bill will be sent to the authentic subscriber. Identity theft could lead to impersonation. This, for example, is easily possible in GSM, as explained in Chapter 14. In GSM, there is lack of mutual authentication, and there is no form of security available within the network itself. Such attacks were possible in the original security solution 802.16 (see chapter 13). This is also possible in TLS using certificates (see Chapter 7) and WEP-based 802.11 solutions. (See Chapter 12.)
7	Man-in-the-middle attack	Impersonation of both ends of communication is a man-in-the-middle attack. For example, this happens in EAPoL (see Chapter 10) where the intruder impersonates an access point and, after getting the session key, acts as a station, thus leading to an attack. This kind of attack is possible at various protocol layers, like application, link layer, and transport layer. Such attacks were possible in original security solution 802.16 (see chapter 13) due to the lack of mutual authentication. In TLS, too, man-in-the-middle attacks are possible (see Chapter 7) and in 802.11 (see Chapter 12).
8	Ping flood	ICMP echo requests are used here to flood a network. This is the earliest form of DoS attack, and several tools are available on the Internet. Filtering packets and stopping the traffic completely from a given address is a common way of dealing with such attacks.
9	Ping of death	IP allows packets of length 65,535 bytes; in ping of death attacks an intruder sends packets larger than this in fragments. The receiving side tries to assemble it, but the large packet causes buffer overflow, thus causing an error. This leads to a crash or reboot of the target. There are patches available for these attacks nowadays.
10	Ping sweep	This ping attack makes use of sending ICMP echo requests to a range of IP addresses and collecting information about the. Tools like Fping and Gping are freely available in the Internet. Alternatives are ICMP time stamp and address mass requests.
11	Routing information protocol (RIP) spoofing	RIP is a routing protocol used for sharing routing table information between routers on large internetworks. This attack causes modifications of routing tables by forging RIP packets.
12	Smurf attack	This is another ICMP echo request attack in which a spoofed ICMP request packet is broadcasted by an intruder in a network. If there are enough hosts, this will cause a lot of traffic in the network, thus causing DoS attacks or degradation of services. In mobile networks, the effect is huge; this again requires better planning of the network and infrastructure security consideration.
13	Sniffing	This basically requires capturing and analyzing network traffic. Sniffing and at the same time attacking the confidentiality of the traffic is possible in GSM (see Chapter 14), the original 802.16 security solution (see Chapter 13), and Bluetooth (see Chapter 11).

Table 16.6 (continued)

ID	Threat	Description
14	Spoofing	With this method, one can make the transmission to appear to be coming from someone other than the one who originated it. Intruders make use of IP address spoofing to falsify source address; ARP spoofing is used to falsify MAC addresses; and domain name server spoofing basically causes domain name server information falsification. This attack could be an issue for mobile networks. Planning of networks should be such that legitimate subscribers do not have access to resources other than what they are authorized for.
15	Teardrop	This attack makes use of the fact that the receiving end receiving duplicate fragments can get into error mode. In such an attack, a teardrop sends two UDP packets with a spoofed source address at the domain name server port, frequently open in firewalls, to a target. On reception of such packets, the target caused memory violation and hung or crashed.
16	DoS	This is a kind of attack in which service cannot be provided or accessed by a network or device. Teardrop is an example for it. Such attacks are possible in all networks and systems in different ways. Such attacks are also possible on TLS. (See Chapter 7.)

16.4 Protections Means

Vendors are starting to port the security products conceived for desktop computers to handheld devices. Once users have identified the risks they need to be protected against, they can choose which product is the best fit. An antivirus is the basic security product that should be installed. The drawback of an antivirus is that it must be updated on a regular basis and needs Internet access to download updates against the latest attacks. To avoid attacks by a new virus, a firewall should be installed to block suspicious data flow.

Table 16.7 reports protection means.

Table 16.7
Security Protection Means

ID	Protection	Description
1	Antispam	Antispam is software that blocks unsolicited e-mail.
2	Antispyware	Antispyware software protects against spyware, browser hijackers, tracking cookies, and other Internet parasites that monitor users' Internet use.

Table 16.7 (continued)

ID	Protection	Description
3	Antivirus	Antivirus software shields against known worms, viruses, and Trojans by comparing potential malware against databases of previously identified malware.
4	Authentication product	This ensures that only legitimate users have access to the device content. If password authentication applies, the number of allowed guesses should be restricted. Two-factor authentication is a preferred solution.
5	Encryption product	This is used for data protection in case a malicious user gains access to the device's resources.
6	Firewall	Firewalls are hardware devices or software applications that control incoming and outgoing network flow. Firewalls can block unauthorized access to a private network but cannot remove malware. (See Chapter 5.)
7	Router	Routers are hardware devices or software applications that route data packets among networks. Routers block unsolicited incoming communications, but they do not protect against most malware. (See Chapter 5.)
8	VPN	VPN creates a tunnel to transmit secure information between two hosts on the network. Leading PDA VPNs exist based on SSL and IPSec. VPN use is suggested also to make up for wireless transmission vulnerabilities. (See Chapter 8.)
9	Wireless security features	When available, they should be deployed. (See Chapters 11–14.)

References

[1] Viega, J., and G. McGraw, *Building Secure Software,* Reading, MA: Addison-Wesley, 2001.

[2] F-secure Virus Descriptions, http://www.f-secure.com/v-descs/cabir.shtml.

[3] Benoit, O., et al., "Mobile Terminal Security," Cryptology e-Print Report 2004/158.

[4] Tulloch, M., *Microsoft Encyclopedia of Security,* Seattle, WA: Microsoft Press, 2003.

List of Acronyms

2G	Second generation
3DES	Triple data encryption standard
3G	Third generation
3GPP	Third Generation Partnership Project
4G	Fourth generation
AAA	Authentication, Authorization, and Accounting
AAAB	AAA broker
AAAF	AAA foreign
AAAH	AAA home
AAD	Additional authentication data
AAS	Advanced Antenna System
AC	Access controller
ACL	Access control list
ACO	Authenticated ciphering offset
AES	Advanced encryption system
AH	Authentication header
AIK	Attestation identity key

AK	Access key
AKA	Authentication and key agreement
AKMP	Authentication and key management protocol
AL	Assurance level
AMPS	Advanced mobile phone system
ANSI	American National Standards Institute
AP	Access point
API	Application programming interface
AR	Access router
ARP	Address resolution protocol
ARPU	Average revenue per unit
ASIC	Application specific integrated circuits
AuC	Authentication center
AVP	Attribute value pair
B3G	Beyond 3G
BIOS	Basic input output system
BLKCNT	Block counter
BS	Base station
CA	Certification authority
CAC	Call (or connection) administration control
CAP	Certified access point
CAPWAP	Control and provisioning of wireless access point
CBC	Cipher block chaining
CBC-MAC	Cipher block chaining message authentication code
CC	Common Criteria
CCEVS	Common Criteria Evaluation and Validation Scheme
CCM	CTR CBC-MAC

CCMP	CCM protocol
CCTL	Common Criteria Testing Laboratory
CDMA	Code division multiple access
CEK	Content encryption key
CESG	Communications Electronics Security Group
CHAP	Challenge handshake protocol
CID	Connection ID
CISP	Cardholder Information Security Program
CLDC	Connected limited device configuration
CLR	Common language runtime
CMV	Cryptographic Module Validation
CN	Corresponding node
CoA	Care of address
COF	Ciphering offset number
CPS	Common part sublayer
CPU	Central processing unit
CRC	Cyclic redundancy check
CRL	Certificate revocation list
CRTM	Core root of trust for measurement
CS	Convergence sublayer
CSP	Cryptographic service provider
CTR	Counter
DDoS	Distributed denial of service
DH	Diffie-Hellman
DHCP	Dynamic host configuration protocol
DMA	Direct memory access
DMZ	Demilitarized zone

DOCSIS	Data over cable service interface specification
DoS	Denial of service
DPA	Differential power analysis
DRM	Digital rights management
DSL	Digital subscriber line
EAL	Evaluation assurance level
EAP	Extensible authentication protocol
EAP-AKA	EAP UMTS authentication and key agreement
EAP-SIM	EAP-enhanced GSM authentication
EAPOL	EAP over LAN
EAP-TLS	EAP transport
EAP-TTLS	EAP tunneled TLS
EAR	Edge access router
EC	European Commission
ECB	Electronic code book (mode)
ECC	Elliptic curve cryptography
EDR	Enhanced data rate
EEPROM	Electrically erasable programmable read-only memory
EK	Endorsement key
EMV	Europay, Mastercard, Visa
ESP	Encapsulating security payload
ETSI	European Telecommunications Standards Institute
FA	Foreign agent
FAST	Flexible authentication via secure tunneling
FCC	Federal Communications Commission
FIP	Fair Information Principles
FIPS	Federal Information Processing Standard

FTP	File transfer protocol
FWA	Fixed wireless access
GFSK	Gaussian frequency shift keying
GMK	Group master key
GPRS	General packet radio system
GSM	Global system for mobile communications
GTK	Group transient key
GW	Gateway
HA	Home address
HAP	High-Altitude Platform
HLR	Home location register
HMAC	Hashed message authentication code
HTTP	Hypertext transport protocol
HTTPS	HTTP over SSL or TLS
IBSS	Independent basic service set
ICAO	International Civil Aviation Organization
ICMP	Internet control message protocol
ICV	Integrity check value
ID	Identity
IDS	Intrusion detection system
IE	Information element
IEEE	Institute of Electrical and Electronics Engineers
IETF	Internet Engineering Task Force
IKE	Internet key exchange
IMEI	International mobile equipment identity
IMS	IP multimedia subsystem
IMSI	International mobile subscriber identity

INIT	Initialization and frame-independent counter
IP	Internet protocol
IPL	Initial program loader
IPSec	Internet protocol security
IRTF	Internet Research Task Force
ISAKMP	Internet security association and key management protocol
ISO	International Standards Organization
ISP	Internet service provider
IT	Information technology
ITSec	Information technology Security Evaluation and Certification
ITU	International Telecommunications Union
IV	Initialization vector
J2ME	Java 2 platform, micro edition
JDK	Java development kit
KCK	Key confirmation key
KEK	Key encryption key
KM	Key modifier
KSSL	Kilobyte SSL
L2CAP	Logical link control and adaptation layer protocol
L2TP	Layer 2 tunneling protocol
LAN	Local area network
LDAP	Lightweight directory access protocol
LED	Light emitting diode
LFSR	Linear feedback shift registers
LLC	Logical Link Control
LM	Link manager
LMP	Link manager protocol

LOS	Line of sight
LTE	Long-term evolution
MAC	Message authentication code
MAC	Medium access control
MAGNET	My Personal Adaptive Global NET
MAN	Metropolitan area network
MAP	Mobile application part
MAPsec	Mobile application part security
MB-OFDM	Multiband OFDM
MBWA	Mobile broadband wireless access
MC	Multicarrier
MD	Message digest
MIC	Message integrity check
MIDP	Mobile information device profile
MIMO	Multiple input multiple output
MIP	Mobile Internet protocol
MK	Master key
MMS	Multimedia messaging subsystem
MMU	Memory management unit
MN	Mobile node
MNO	Mobile network operator
MPDU	MAC protocol data unit
MSC	Message Sequence Chart
MSDU	MAC service data unit
MSS	Mobile SS
MT	Mobile terminal
NAI	Network address identifier

NAPT	Network address port translation
NAS	Network access server
NAT	Network address translator
NATO	North Atlantic Treaty Organization
NAV	Network allocation vector
NIAP	National Information Assurance Partnership
NIST	National Institute for Standards and Technology
NLOS	Non-LOS
NSA	National Security Agency
OCSP	Online certificate status protocol
ODRL	Open digital rights language
OECD	Organization of Electronic Cooperation and Development
OFDM	Orthogonal frequency division multiplexing
OMA	Open Mobile Alliance
OP	Open platform
OS	Operating system
OSI	Open system interconnection
OTP	One-time password
PAC	Protected access credential
PAK	Primary AK
PAP	Password authentication protocol
PC	Personal computer
PCI	Payment card industry
PCR	Platform configuration registers
PDA	Personal digital assistant
PDG	Packet data gateway
PDSN	Packet data serving node

PDU	Protocol data unit
PEAP	Protected EAP
PHY	Physical layer
PIN	Personal identification number
PKI	Public key infrastructure
PKM	Privacy key management
PLMN	Public land mobile network
PLMS	Public land mobile stations
PMK	Pairwise master key
PN	Packet number
PP	Protection profile
PPP	Point-to-point protocol
PPSC	Privacy Protection Study Commission
PPTP	Point-to-point tunneling protocol
PRF	Pseudorandom function
PRN	Pseudorandom number
PRNG	Pseudorandom number generator
PS	padded string
PSH	Payload Suppression Header
PSK	Preshared key
PSK	Pairwise shared key
PSTN	Public switched telephone network
PTK	Pairwise transient key
PWLAN	Public WLAN
QoS	Quality of service
RADIUS	Remote authentication dial-in user service
RAN	Radio access network

RAM	Random access memory
RF	Radio frequency
RFC	Request for comment
RIP	Routing information protocol
RNSN	Radio network serving node
RRM	Radio resource management
RSA	Rivest-Shamir-Adlemann
RSN	Robust security network
RSNA	Robust security network association
RTP	Real-time protocol
SA	Security association
SAD	Security association database
SADB	Security association database
SAE	System architecture evolution
SAID	Security association identifier
SAML	Security authorization markup language
SAP	Service access point
SAS	Security Accreditation Scheme
SDR	Software defined radio
SDU	Service data unit
SEG	Security Gateway
SET	Secure electronic transaction
SFID	Service flow identifier
SHA	Secure hash algorithm
SIG	Special interest group
SIM	Subscriber identity module
SIP	Session identity protocol

SKEME	Secure key exchange mechanism
SLA	Service level agreement
SMIME	Secure/multipurpose Internet mail extensions
SMS	Short message service
SNMP	Secure network management protocol
SOI	Son of IKE
SoC	System on a chip
SP	Security policy
SPA	Simple power analysis
SPD	Security policy database
SPD	Security policy database
SPI	Security parameter index
SRES	Signed response
SRK	Storage root key
SS	Subscriber station
SS7	Signaling system number 7
SSH	Secure shell
SSL	Secure socket layer
ST	Security target
STA	Station
STS	Station to station
TA	Transmission address
TACACS	Terminal access controller access control system
TACS	Total access communication system
TAP	Transferred account procedure
TCG	Trusted Computing Group
TCP	Transport control protocol

TCP/IP	Transport control protocol/Internet protocol
TDMA	Time division multiple access
TEK	Traffic encryption key
TEMPEST	Telecommunications Electronics Material Protected from Emanating Spurious Transmissions
TK	Temporal key
TKIP	Temporal key Internet protocol
TLS	Transport layer security
TMSI	Temporary mobile subscriber identity
TOE	Target of evaluation
TPM	Trusted Platform Module
TSA	Time stamp authority
TSN	Transitional security network
TSC	TKIP sequence counter
TSS	Trusted platform support service
TTLS	Tunneled TLS
TTP	Trusted third party
UAM	Universal access method
UART	Universal asynchronous receiver-transmitter
UbiCom	Ubiquitous communication
Ubicomp	Ubiquitous computing
UDP	User datagram protocol
UE	User equipment
UMTS	Universal mobile telecommunications system
USB	Universal serial bus
USIM	Universal subscriber identity module
USIM	UMTS SIM
USM	User-based security model

UWB	Ultrawideband
VoIP	Voice over IP
VLR	Visited local register
VPN	Virtual private network
WAP	Wireless access protocol
WAS	Wireless access server
WCDMA	Wireless code division multiple access
WEP	Wired equivalent protocol
WG	Working group
WiBro	Wireless broadband
WIM	WAP identity module
WiMAX	Worldwide interoperability microwave access
WISP	Wireless Internet Service Provider
WISPr	WISP Roaming
WLAN	Wireless local area network
WMAN	Wireless metropolitan area network
WPA	WiFi protected area
WPAN	Wireless personal area network
WS	Web services
WSS	WS security
WTLS	Wireless TLS
WTP	Wireless trusted platform
WTP	Wireless termination point
WWAN	Wireless wide area network
XML	Extensible markup language
X-DoS	XML denial of service

About the Authors

Selim Aissi is an architecture manager in Intel's Mobile Platform Group, where he leads the architecture of manageability, security, and wireless connectivity technologies for next generation mobile platforms. During his six years at Intel, Selim has held several management and senior architecture roles, both in research and product development. During this tenure, he has developed technologies for wireless cellular processors (Intel XScale and PXA), mobile platforms (Centrino Mobile Technology), and several software and protocol stacks. He has also represented Intel in 3GPP SA3, IEEE, Bluetooth SIG, UDDI, and other standards development organizations. Before joining Intel in 2000, he worked at the University of Michigan, General Dynamics' M1A2 Battlefield Tank Division, General Motors' Embedded Controller Excellence Center, and Applied Dynamics International. Selim has served on the review board of several publications and conferences, including ACM CCS, ACM SWS, and the advisory board of several conferences and publications, and he is the vice-chair of the Security and Management (SAM) Conference. He holds a Ph.D. in aerospace engineering from the University of Michigan and is a senior member of the IEEE.

Nora Dabbous holds a Master in Telecommunications Engineering from the University of Bologna, Italy. She has worked for over seven years in computer security, both in research and development and on product security in the Security Labs at Gemalto, France. Initially, Nora studied side channel attacks on smart cards and developed countermeasures. Now she is a security architect focusing on product security in the telecommunications, banking, and digital rights management fields. She has participated in standardization efforts to specify security requirements for mobile platforms at OMTP, the Open

Mobile Terminal Platform organization. She has led work packages in multiple European projects under the Information Society Technologies Program and Microelectronics Developments for European Applications, including Trust-ES and Mosquito. She has served on the review board of several conferences and has several patent applications in the smart card security field.

Anand Raghawa Prasad, Senior Member, IEEE, leads the network-level security model group as a manager at DoCoMo Euro-Labs, Munich, Germany. He was born in Ranchi, India. He received his Ph.D. degree in the field of WLANs: protocols, security, and deployment, and MSc (Ir.) degree in Similarity in ATM Network Traffic from Delft University of Technology, the Netherlands, in 2003 and 1996, respectively. From 1996 to 1998 he worked as research engineer and, later, project leader at Uniden Corporation, Tokyo, Japan. From 1998 to 2000, he was a distinguished member of the technical staff and worked as systems architect for IEEE 802.11-based WLANs (WaveLAN and ORiNOCO) at Lucent Technologies, Nieuwegein, the Netherlands. Subsequently, he was technical director at Genista Corporation, Tokyo, Japan, and technical advisor of PCOM:I^3, Aalborg, Denmark. He was a voting member of IEEE 802.11, and he is an active participant of 3GPP. In addition to his publications in peer reviewed journals, international conferences, and chapters in books, he has 25 patent applications in the field of wireless communications and networking and has coedited the book *WLAN Systems and Wireless IP for Next Generation* and coauthored the book *802.11 WLANs and IP Networking: Security QoS and Mobility,* both published by Artech House, in 2002 and 2005, respectively. He has participated in the organization of several international conferences, including VTC, PIMRC, ACM WMASH, and WPMC. He is Co-General chair of ACM-Multimedia. He was a guest editor of a special issue on "Security for Next Generation Communications," the Kluwer International Journal on *Wireless Personal Communications,* and he is a guest editor of a special issue on "Seamless Handover in Next Generation Wireless/Mobile Communications," The Springer International jounral on *Wireless Personal Communications.*

Index

The Artech House Universal Personal Communications Series

Ramjee Prasad, Series Editor

Towards the Wireless Information Society: Systems, Services, and Applications, Ramjee Prasad, editor

Universal Wireless Personal Communications, Ramjee Prasad

WCDMA: Towards IP Mobility and Mobile Internet, Tero Ojanperä and Ramjee Prasad, editors

Wideband CDMA for Third Generation Mobile Communications, Tero Ojanperä and Ramjee Prasad, editors

Wireless Communications Security, Hideki Imai, Mohammad Ghulam Rahman and Kazukuni Kobara

Wireless IP and Building the Mobile Internet, Sudhir Dixit and Ramjee Prasad, editors

WLAN Systems and Wireless IP for Next Generation Communications, Neeli Prasad and Anand Prasad, editors

WLANs and WPANs towards 4G Wireless, Ramjee Prasad and Luis Muñoz

For further information on these and other Artech House titles, including previously considered out-of-print books now available through our In-Print-Forever® (IPF®) program, contact:

Artech House
685 Canton Street
Norwood, MA 02062
Phone: 781-769-9750
Fax: 781-769-6334
e-mail: artech@artechhouse.com

Artech House
46 Gillingham Street
London SW1V 1AH UK
Phone: +44 (0)20 7596-8750
Fax: +44 (0)20 7630-0166
e-mail: artech-uk@artechhouse.com

Find us on the World Wide Web at: www.artechhouse.com